WORKSHOP PHYSICS® ACTIVITY GUIDE

Activity-Based Learning

MODULE 4: ELECTRICITY AND MAGNETISM

Electrostatics, DC Circuits,
Electronics, and Magnetism
(Units 19-27)

PRISCILLA W. LAWS
DICKINSON COLLEGE

with contributing authors:
ROBERT J. BOYLE
PATRICK J. COONEY
KENNETH L. LAWS
JOHN W. LUETZELSCHWAB
DAVID R. SOKOLOFF
RONALD K. THORNTON

JOHN WILEY & SONS, INC.

Cover Image: James Fraher/Image Bank/Getty Images

To order books or for customer service, please call 1-800-CALL-WILEY (225-5945).

ISBN 978-0-471-64116-2

Printed in the United States of America

SKY10042717_021023

Printed and bound by Quad/Graphics

CONTENTS

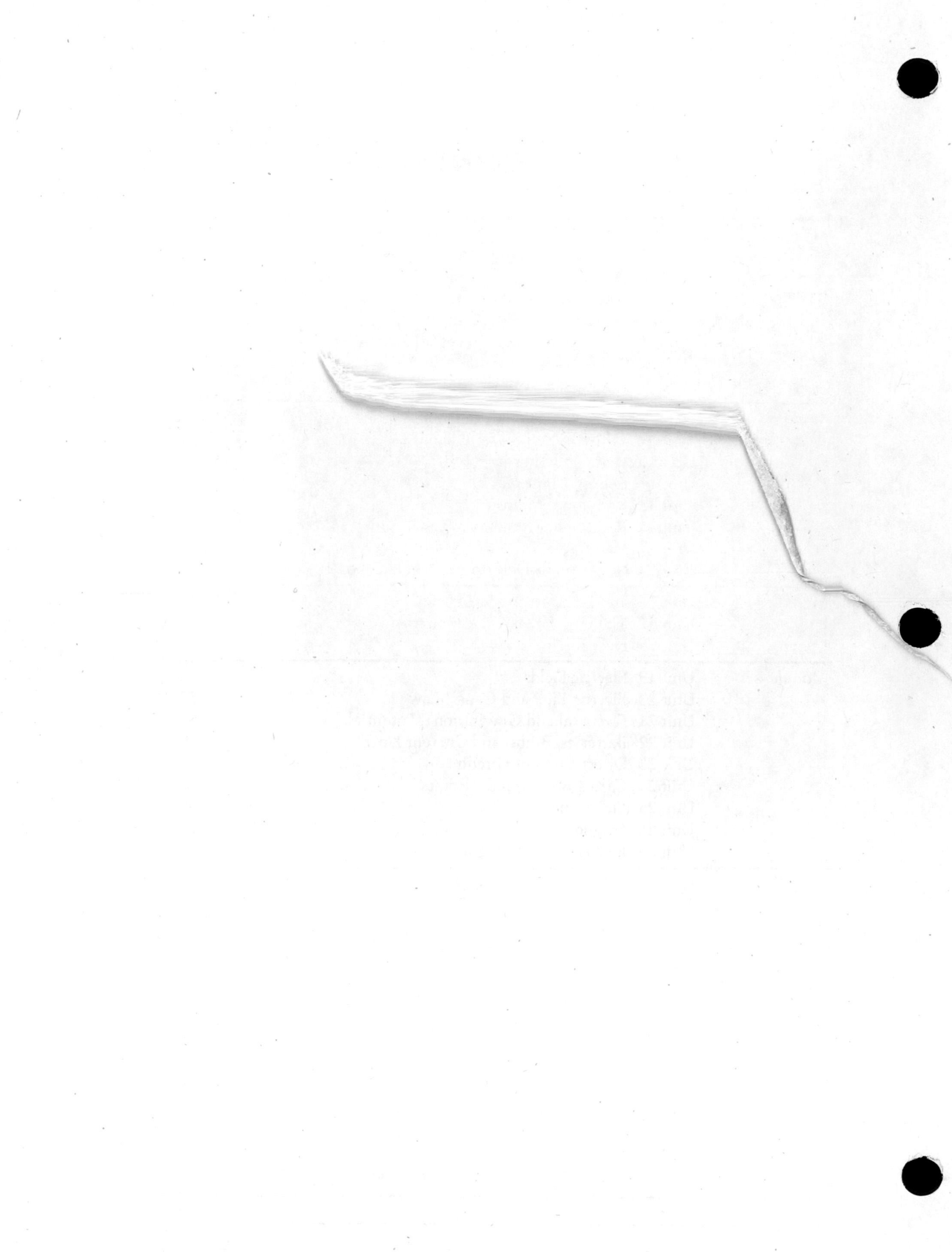

Name ______________________ Section ____________ Date ____________

UNIT 19: ELECTRIC FIELDS

While this physicist is touching a Van de Graaff generator, his hair is sticking out. What is this generator doing that makes his hair stand on end? After you study the fundamental nature of electrical forces in this unit, you should be able to answer this question.

UNIT 19: ELECTRIC FIELDS

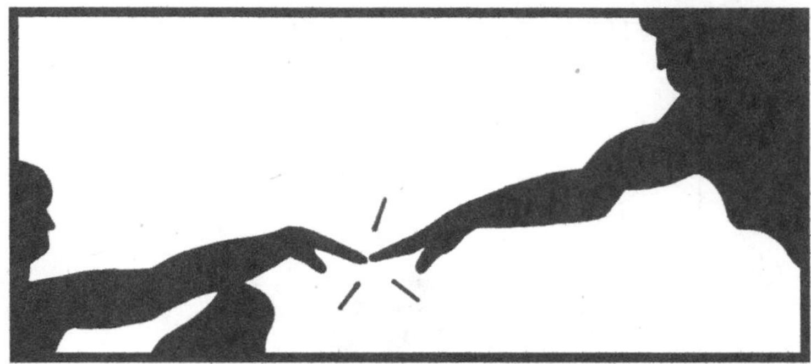

Electricity is a quality universally expanded in all the matter we know, and which influences the mechanism of the universe far more than we think.

Charles Dufay (1698–1739)

OBJECTIVES

1. To discover some of the basic properties of particles that carry electric charges.
2. To understand how Coulomb's law describes the forces between charged objects.
3. To understand the concept of electric fields.
4. To learn how to calculate the electric field associated with charges that are distributed throughout an object.

Fig. 19.1.

22.1. OVERVIEW

On cold, clear days, rubbing almost any object seems to cause it to be attracted to or repelled from other objects. After being used, a plastic comb will pick up bits of paper, hair, and cork, and people wearing polyester clothing in the winter walk around cursing the phenomenon dubbed in TV advertisements as "static cling." We are going to begin a study of electrical phenomena by exploring the nature of the forces between objects that have been rubbed or that have come into contact with objects that have been rubbed. These forces are attributed to a fundamental property of the constituents of atoms known as charge. The forces between particles that are not moving or that are moving relatively slowly are known as *electrostatic forces*.

We start our study in the first several sections by exploring the circumstances under which electrostatic forces are attractive and under which they are repulsive. This should allow you to determine how many types of charge there are. Then we can proceed to a qualitative study of how the forces between charged objects depend on the amount of charge the objects carry and on the distance between them. This will lead to a formulation of *Coulomb's law*, which expresses the mathematical relationship of the vector force between two small charged objects in terms of both distance and quantity of charge. In several later activities you will be asked to verify Coulomb's law quantitatively by performing a video analysis of the repulsion between two charged objects as they get closer and closer together.

Finally, at the end of the unit we will define a quantity called *electric field* that can be used to determine the net force on a small test charge due to the presence of other charges. You will then use Coulomb's law to calculate the electric field at various points of interest arising from some simply shaped charged objects.

ELECTROSTATIC FORCES

19.2. EXPLORING THE NATURE OF ELECTRICAL INTERACTIONS

You can investigate some properties of electrical interactions with the following equipment. Each student should have:

- 4 Scotch tapes, approx. 10 cm long
- 2 small rod stands
- 2 threaded Styrofoam balls (with low mass)
- 2 threaded, metal-coated Styrofoam balls (with low mass)
- 1 hard plastic rod
- 1 fur
- 1 glass rod
- 1 polyester cloth
- 2 metal rods
- 2 right angle clamps

Recommended group size:	2	Interactive demo OK?:	Y

The nature of electrical interactions is not obvious without careful experimentation and reasoning. We will first state two hypotheses about electrical interactions. We will then observe some electrical interactions and determine whether our observations are consistent with these hypotheses.

Hypothesis One: The interaction between objects that have been rubbed is due to a *property* of matter that we will call *charge*.* There are *two* types of electrical charge that we will call, for the sake of convenience, positive charge and negative charge.

Hypothesis Two: Excess charge moves readily on certain materials, known as conductors, and not on others, known as insulators. In general, metals are good conductors, while glass, rubber, and plastic tend to be insulators.

Note: In completing the following activities, you are not allowed to state results that you have memorized previously. You must devise a sound and logical set of reasons to support the hypotheses.

Hypothesis One: Testing for Different Types of Charge

Try the following suggested activities. Mess around and see if you can design careful, logical procedures to demonstrate that there are at least two types of charge. Carefully explain your observations and reasons for any conclusions

* A property of matter is not the same thing as the matter itself. For instance, a full balloon has several properties at once—it can be made of rubber or plastic, have the color yellow or blue, have a certain surface area and so on. Thus, we don't think of charge as a substance but rather as a property that certain substances can have at times. It is easy when speaking and writing casually to refer to excess charge as if it were a substance. Don't be misled by this practice that we will all indulge in at times during the next few units.

you draw. **Hint:** What procedures should you use to generate two objects that carry the same type of excess charge?

Fig. 19.2.

19.2.1. Activity: Interactions of Scotch Tape Strips

a. You and your partner should each place a 10 cm or so strip of Scotch tape on the lab table with the sticky side down. The end of each tape should be curled over to make a non-stick handle. Peel your tape off the table and bring the non-sticky side of the tape toward your partner's strip. What happens? How does the distance between the tapes affect the interaction between them?

b. Place two strips of tape on the table sticky side down and label them "B" for bottom. Press another strip of tape on top of each of the B pieces; label these strips "T" for top. Pull each pair of strips off the table. Then pull the top and bottom strips apart.

1. Describe the interaction between two top strips when they are brought toward one another.

2. Describe the interaction between two bottom strips.

3. Describe the interaction between a top and a bottom strip.

c. Are your observations of the tape strip interactions consistent with the hypothesis that there are two types of charge? Please explain your answer carefully, in complete sentences, and cite the outcomes of *all* your observations.

In earlier times scientists charged objects by rubbing a rubber rod with fur or by rubbing a glass rod with silk. These days we use polyester instead of silk and hard plastic instead of rubber. In the next activity you can get more experience studying the interactions between charged objects using techniques developed by early investigators. **Note:** In the activities that follow, your observations will not be valid if you touch the balls with your hands after charging them.

19.2.2. Activity: Charging Styrofoam Balls with Rods

a. Try rubbing a black plastic rod with fur and then use the rod to touch a pair of small Styrofoam balls hanging from nonconducting threads. What happens to the hanging balls? What happens when you bring the plastic rod near the balls?

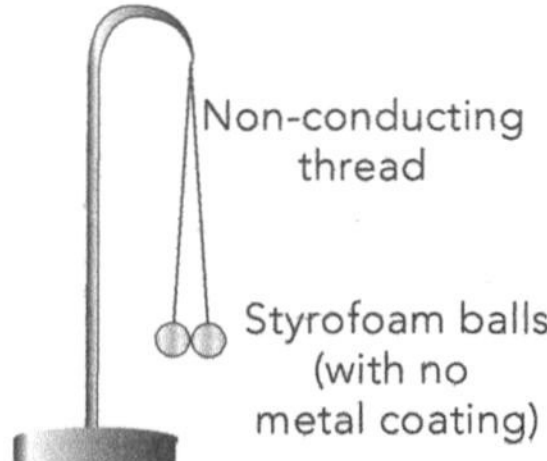

Fig. 19.3.

b. What happens if you rub a glass rod with polyester and then bring it into the vicinity of the balls that were charged with the plastic rod?

Benjamin Franklin

c. Recalling the interactions between like and unlike charged objects that you observed before, can you explain your observations?

d. Touch the entire surface of each of the two charged Styrofoam balls with your hands. Now what happens when you let them hang again? Is there an interaction between them?

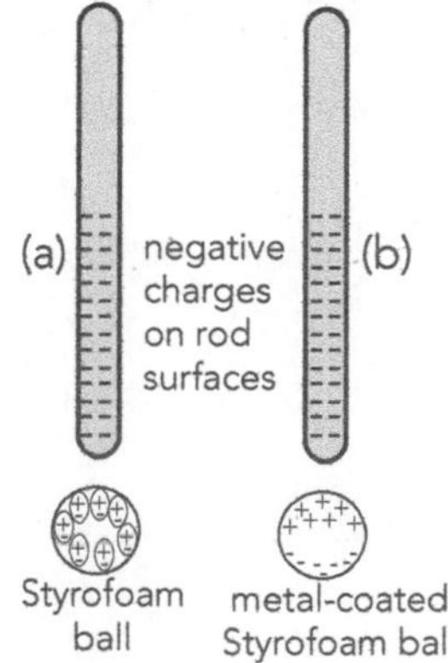

Fig. 19.4. Depiction of how excess electrons can influence a neutral insulator and a neutral conductor. (a) The electrons in the Styrofoam insulator are repelled from those in the rod but stay with their atoms. (b) Electrons in a metal-coated ball are repelled by the electrons in the rod and are free to move as far away from the rod as possible.

Benjamin Franklin *arbitrarily* assigned the term "negative" to the nature of the charge that results when a hard plastic rod (or, in his day, a rubber rod) is rubbed with fur. Conversely, the nature of the charge found on the glass rod after it is rubbed with silk is defined as "positive." (The term "negative" could just as well have been assigned to the charge on the glass rod; the choice was purely arbitrary.)

Hypothesis Two: Using Induction to Distinguish Conductors from Nonconductors

Scientists believe that most matter is made of atoms that contain positive and negative charges associated with protons and electrons respectively. When electrons in an atom surround an equal number of protons the charges neutralize each other and the atom does not interact with other charges outside the solid. In some types of solid materials, known as insulators, the electrons are tightly bound to the protons in the atoms and do not move away from their atoms. However, in other solids known as conductors, the electrons but not the protons are free to move under the influence of other charges. The process by which external charges can cause electrons in an object to rearrange themselves so that the positive and negative charges no longer neutralize each other is know as *induction*.

In the next activity you will study an interaction that involves induction and one that doesn't. You should observe the interaction between a uncharged non-metal-coated Styrofoam ball that is hanging freely and a charged plastic rod. You should also observe the interaction between the charged rod and an uncharged metal-coated Styrofoam ball that is hanging. Note: Before making each observation, touch each of your hanging balls to make sure they are not charged.

19.2.3. Activity: Insulators, Conductors, and Induction

a. What happens when you bring a negatively-charged plastic rod near an uncharged insulator consisting of a single Styrofoam ball ?

1. Before the two objects touch:

2. After they touch:

b. Repeat observation a. using an uncharged metal-coated Styrofoam ball.

1. Before the two objects touch:

2. After they touch:

c. Use Hypothesis Two, which claims that charges move readily on conductors, to explain why the metal-covered ball is *attracted* to the rod *before* touching and *repelled after* touching it.

d. Use Hypothesis Two, which claims that if an object is an insulator, its electrons will stay in their vicinity of their atoms, to explain why the uncoated Styrofoam ball is still attracted to the rod after touching it.

Note: We picture "uncharged" objects made up of a huge number of atoms having an equal number of negatively charged electrons that swarm around positively charged protons in the atomic nucleii. A "charged" object has either more electrons than protons or fewer electrons than protons. For this reason, we refer to a charged object as having "excess charge" or "a net charge."

19.3. FORCES BETWEEN CHARGES– COULOMB'S LAW

Coulomb's law is a mathematical description of the fundamental nature of the electrical forces between charged objects that are either spherical in shape or small compared to the distance between them (so that they act more or less like point particles). This law relates the force between small charged objects to the excess charge on the objects and the distance between them. Coulomb's law is usually stated without experimental proof in most introductory physics textbooks. Instead of just accepting the textbook statement of Coulomb's law, you are going to determine qualitatively how the charge on two objects and their separation affect the mutual force between them. These objects could be, for instance, two metal-coated balls, or perhaps a small metal ball affixed to the tip of an insulated rod and one metal-coated ball. For this set of observations you will need:

- 2 small rod stands
- 2 threaded, metal-coated Styrofoam (or ping pong) balls with low mass
- 1 hard plastic rod
- 1 piece of fur
- 1 glass rod
- 1 polyester cloth
- 2 metal rods
- 2 right angle clamps

Recommended group size:	2	Interactive demo OK?:	N

Note: Coulomb devised a clever trick for determining how much force charged objects exert on each other without knowing the actual amount of charge on the objects. Coulomb transferred an unknown amount of charge, q, to a conductor. He then touched the newly charged conductor to an identical uncharged one. The conducting objects would quickly exchange charge until both had a net or excess charge $q/2$ on them. After observing the effects with $q/2$, Coulomb would discharge one of the conductors by touching a large piece of metal to it and then repeat the procedure to get $q/4$ on each conductor, and so on.

19.3.1. Activity: Dependence of Force on Charge, Distance, and Direction–Qualitative Observations

Consider a pair of conductors, each initially having excess charge $q_A = q_B = q/2$. These conductors are hanging from strings in the configuration shown in Figure 19.5.

Use the following diagrams to sketch what you predict will happen to the angles the charged objects make with respect to the vertical as compared to their initial angles when $q_A = q_B = q/2$. In each case give the reasons for your prediction. Then make the observation and sketch what you observed.

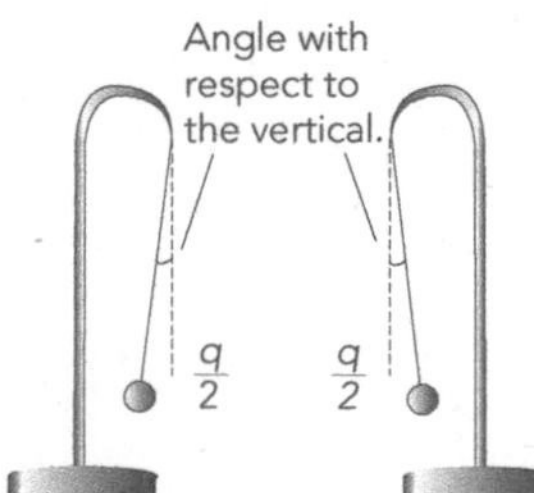

Fig. 19.5. Original position of pivots for 2 charged conductors.

a. What if the charged conductors still each have an excess charge of $q_A = q_B = q/2$ but the pivots for the strings are moved closer together as shown in the diagram below?

Prediction

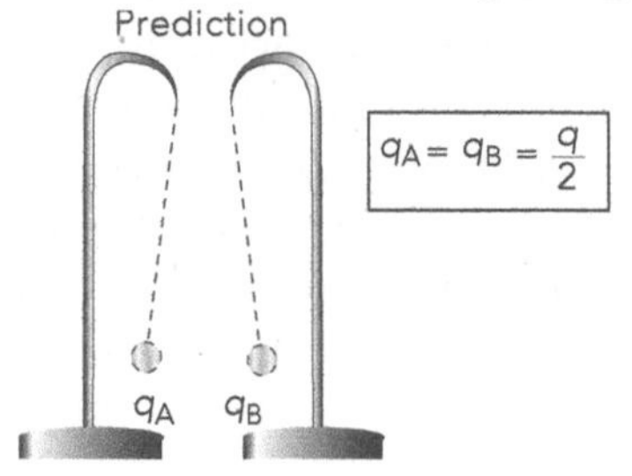

Reasons for prediction.

Observation

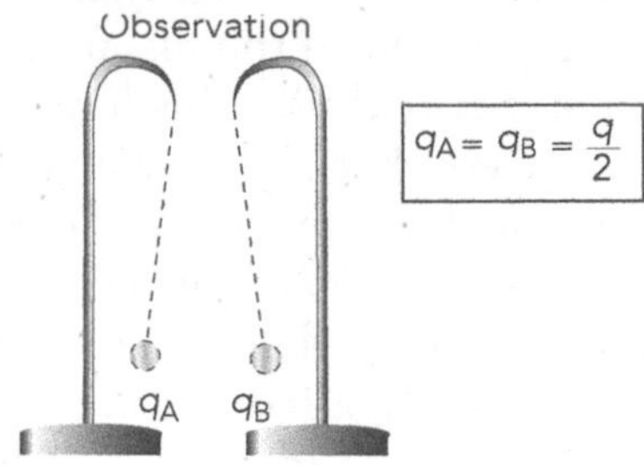

New explanation if observation and prediction disagree.

b. What seems to happen to the force of interaction between the charged conductors as the distance between them decreases?

c. What if the pivots are moved back to their original position as shown in Figure 19.5 but the amount of charge on each conductor is decreased so that $q_A = q_B = q/4$?

Prediction

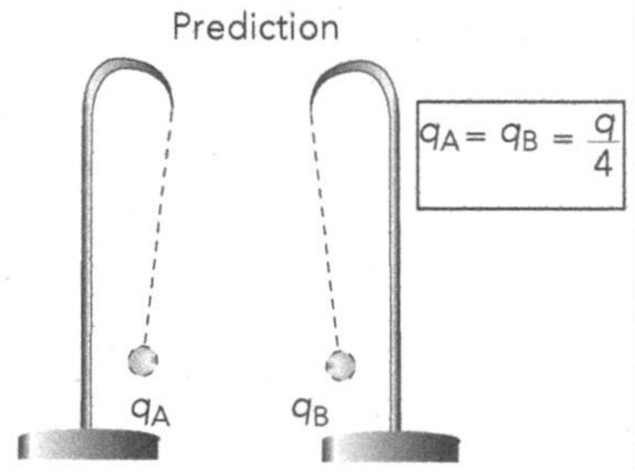

Reasons for prediction.

Observation

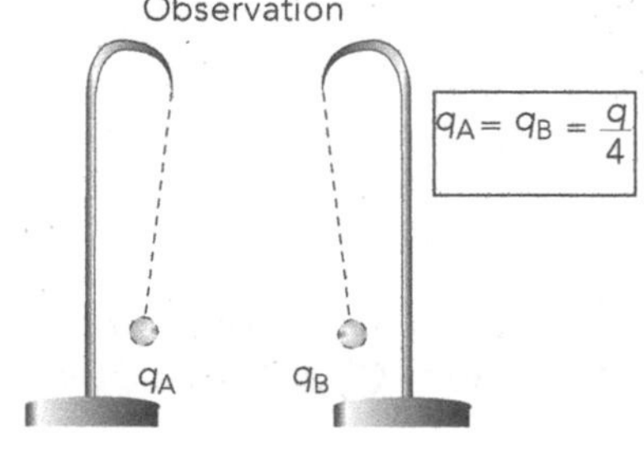

New explanation if observation and prediction disagree.

d. Does the force of interaction between charged objects seem to increase or decrease as the charge decreases?

e. Assume that the locations of the pivots are still held at the original spacing. What if *one* of the conductors q_A still has a charge of $q/4$ while the other one is discharged completely so that $q_B = 0$. The observation may surprise you. Can you explain it? **Hint:** Do Newton's Third Law or the idea of induction come into play?

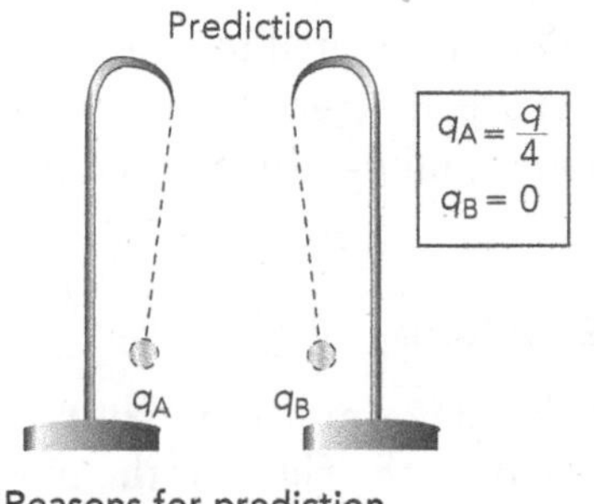

Reasons for prediction.

Observation

$q_A = \frac{q}{4}$

$q_B = 0$

q_A q_B

New explanation if observation and prediction disagree.

f. On the basis of the observations you have already made, explain why the force between the two charged objects seems to lie along a line between them. **Hint:** Could Newton's Third Law hold if the mutual repulsion or attraction forces did not lie on a line between the objects? Explain.

19.4. THE MATHEMATICAL FORMULATION OF COULOMB'S LAW

Coulomb's law asserts that the magnitude of the force between two electrically charged spherical objects is *directly proportional to the product of the amount of excess charge on each object and inversely proportional to the square of the distance* between the centers of the spherical objects. The direction of the force is along a line between the two objects and is attractive if the particles have opposite signs and repulsive if they have like signs. All of this can be expressed by the equation following which represents the electrostatic force exerted on q_A due to q_B.

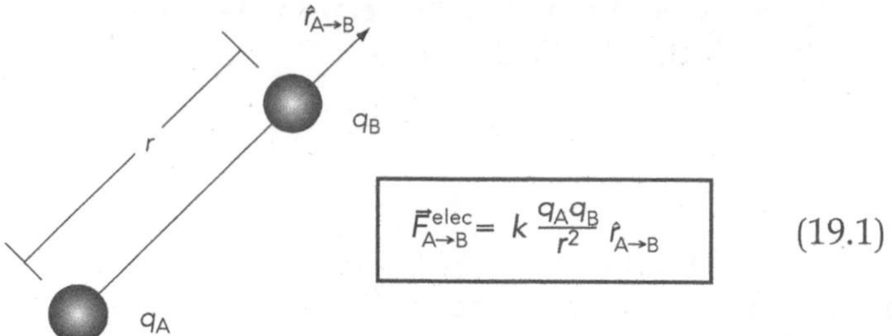

$$\vec{F}^{elec}_{A\to B} = k\frac{q_A q_B}{r^2}\hat{r}_{A\to B} \quad (19.1)$$

Fig. 19.6. Diagram showing the direction of the unit vector $\hat{r}_{A->B}$ used in the Coulomb force equation that describes the influence of charge q_B on charge q_A.

The $\hat{r}_{A->B}$ with a "hat" over it is a unit vector directed from q_A to q_B, r^2 is the square of the distance between the two charged objects in meters, k is the Coulomb constant (9.0×10^9 N·m^2/C^2), and q_A and q_B are the charges in coulombs.

19.4.1. Activity: "Reading" the Coulomb Equation

a. Draw the direction of the unit vector in the diagram below. **Note:** The direction of this vector does not depend on the signs or the magnitudes of the charges.

b. In the table below, indicate the sign of the product of q_A and q_B for each combination of positive and/or negative charges.

Sign of q_A	Sign of q_B	Sign of $q_A \cdot q_B$
+	+	
+	–	
–	+	
–	–	

c. Use an arrow to indicate the direction of the force exerted by q_A on q_B if the charges are both positive or both negative.

d. Use an arrow to indicate the direction of the force exerted by q_A on q_B if the charges have opposite signs (that is, one is positive and one is negative).

e. If the force vector $\vec{F}^{\text{elec}}_{A\to B}$ is in the opposite direction from the unit vector $\hat{r}_{A->B}$, the unit vector must be multiplied by a negative number. Where does this negative number come from in the Coulomb equation? Does this negative number indicate a repulsive force or an attractive force?

f. In the Coulomb equation, does the magnitude of the force decrease as either q_A or q_B decreases? Why or why not?

g. In the Coulomb force equation, does the magnitude of the force increase as the distance between the charged objects decreases? Why or why not?

h. In the diagram below, show the direction of the unit vector that describes the force of q_B on q_A.

i. Is Coulomb's law consistent with Newton's Third Law? In particular, how do $\vec{F}^{\text{elec}}_{A\to B}$ and $\vec{F}^{\text{elec}}_{B\to A}$ compare in magnitude? In direction?

To get some more practice with reading and using the Coulomb's law equation do the following vector calculations. You may need to brush up on vectors!

Reminder: $\hat{\imath}$ and $\hat{\jmath}$ represent unit vectors pointing along the x- and y-axes, respectively.

19.4.2. Activity: Using Coulomb's Law for Calculations

a. Consider two point-like objects with excess charges on them. Suppose they lie along the x-axis. A net charge of 2.0×10^{-9} C is located at $x =$ 3.0 cm and a net charge of -3.0×10^{-9} C point charge is located at $x =$ 5.0 cm. What is the magnitude of the force on the negatively charged object due to the presence of the positively charged object? What is its direction? Express the force as a vector quantity using unit vector notation.

b. Suppose the unit vector $\hat{r}_{A->B}$ makes an angle θ with the x-axis as shown in the following diagram. Use unit vector notation to express $\hat{r}_{A->B}$ in Cartesian coordinates in terms of $\sin\theta$ and $\cos\theta$. **Hint:** $\hat{r}$ signifies a "unit vector" that serves as a non-dimensional pointer (or direction indicator).

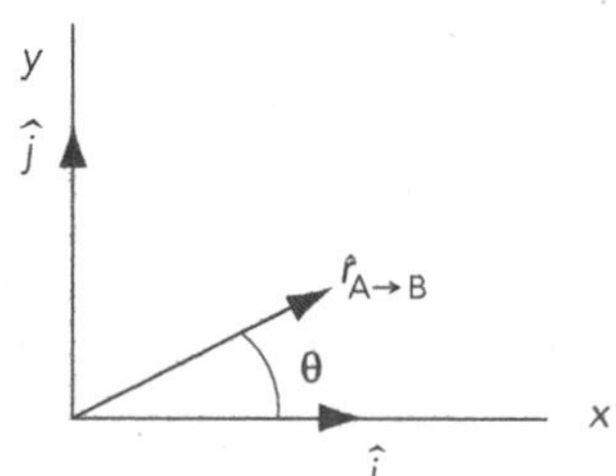

c. Suppose the -3.0×10^{-9} C point charge is moved to $x = 5.0$ cm and $y = 6.0$ cm. What is the magnitude of the force exerted by the negative point charge on the positive point charge? What is its direction? Express the force as a vector quantity using unit vector notation. Then draw a diagram of this situation, indicating the positions of the charges and the force vector. **Hint:** (1) Calculate the magnitude of the force. (2) Figure out what angle the force vector makes with respect to the x-axis. (3) Decompose the force vector into x- and y-components.

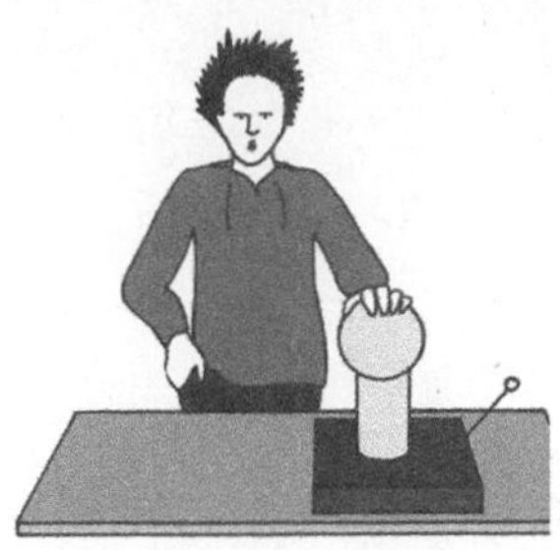

Fig. 19.7.

QUANTITATIVE ASPECTS OF COULOMB'S LAW

19.5. DEMONSTRATION OF ELECTROSTATIC DISCHARGES

In addition to exploring the nature of the relatively small collections of electrical charge that result from rubbing objects together, you can examine two demonstrations involving relatively high levels of electrical charge being "discharged."

The Van de Graaff Generator

Ben Franklin and others recognized that electrical charge can be "produced" by doing mechanical work. The Van de Graaff generator will be demonstrated briefly. This device produces a relatively high density of electrical charge.

Demonstration of the Storm Ball

This popular device has been sold at specialty and gift stores as an amusement. It provides a dramatic demonstration of electrostatic discharge.

For these demonstrations you will need:

- 1 Van de Graaff generator
- 1 storm ball

Recommended group size:	All	Interactive demo OK?:	Y

19.6. QUANTITATIVE VERIFICATION OF COULOMB'S LAW

In the late eighteenth century Charles Coulomb used a torsion balance and a great deal of patience to verify that the force of interaction between small spherical charged objects varied as the inverse square of the distance between them. Verification of the inverse square law can also be attempted using modern apparatus.

A small, conducting sphere can be placed on the end of an insulating rod and can then be charged negatively using a plastic rod that has been rubbed with fur. This charged sphere can be used as a prod to cause another charged sphere, suspended from two threads, to rise to a higher and higher angle as the prod comes closer, as shown in the diagram below. A video camera can be used to record the angle of rise, θ, of the suspended object as well as the distance between the prod and the suspended object.

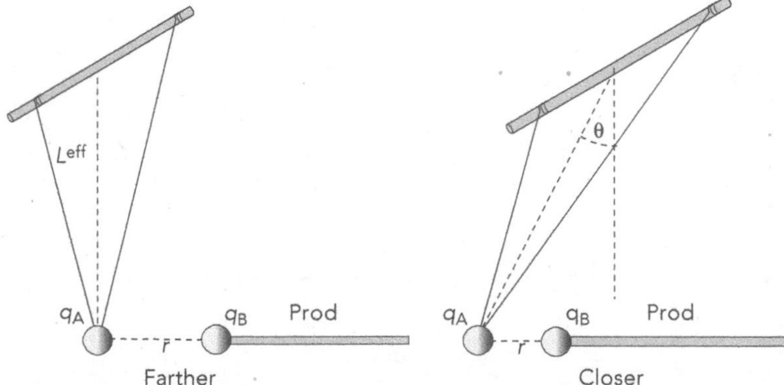

Fig. 19.8. An illustration of how a charged prod might affect the position of a charged conducting ball hanging from a bifilar pendulum.

Using the laws of mechanics, it is possible to determine the relationship between the Coulomb force on the small sphere and the angle through which it rises above a vertical line. Thus, you should be able to measure the Coulomb force on q_B as a function of the distance between q_A and q_B.

Before proceeding with the video analysis, let's take time out to determine the theoretical angle of rise, θ, of a charged sphere of mass m due to a Coulomb force on it. This force is the result of the presence of another charged object that lies in the same horizontal plane as the suspended mass. This situation is shown in the diagram below.

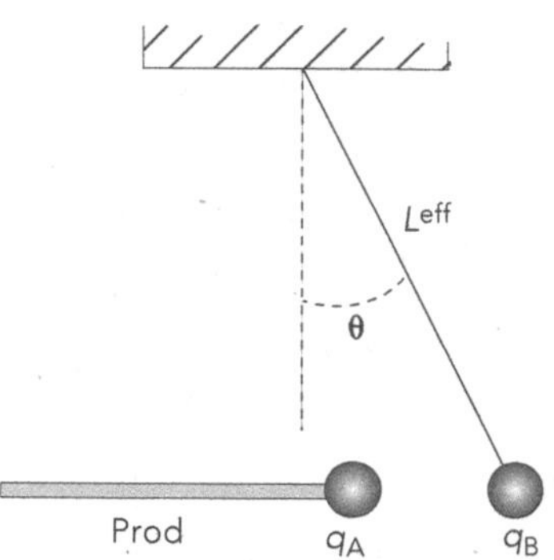

Fig. 19.9. Diagram showing essential information needed to determine free body forces on the charged pendulum bob.

19.6.1. Activity: Forces on a Suspended Charged Object–Theory

a. Draw a vector diagram with arrows showing the direction of each of the forces on the mass m, including the gravitational force, $\vec{F}^{grav}$, the tension in the string, $\vec{F}^{tens}$, and a horizontal electrostatic force due to the charge on the prod, $\vec{F}^{elec}$.

b. Show that, if there is no motion in the vertical direction, and g is the local gravitational constant, then

$$|\vec{F}^{tens}| = \frac{mg}{\cos\theta}$$

c. Show that, if there is no motion along the horizontal direction, then

$$F^{elec} - F^{tens} \sin\theta = 0 \text{ N}.$$

where F^{elec} and F^{tens} represent force magnitudes.

d. Show that $F^{elec} = mg \tan \theta$.

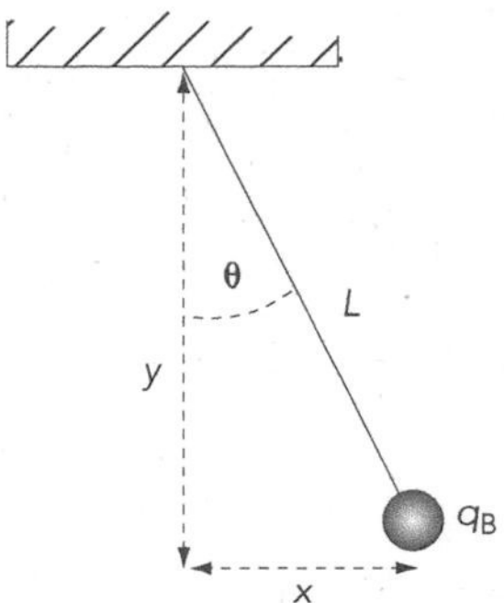

Fig. 19.10.

e. Find $\tan \theta$ as a function of x and L. **Hint:** Start by finding y as a function of x and L.

f. Combine the equations you derived in parts d. and e. to find the equation needed to calculate the experimentally-determined magnitude of the electrostatic force, F^{elec}, as a function of L, x, m, and the local gravitational constant g.

Finally, you can turn to the task of obtaining a videotape of how the mass rises as a function of its horizontal distance from the prod. Then, using the equation you obtained in Activity 19.6.1, you can analyze the video data to determine the magnitude of the electrical force F^{elec} as a function of r, the distance between the centers of the prod and the rising ball. To do this you will need the following apparatus:

- 1 metal-coated, threaded ping pong ball (with low mass)
- 2 non-conducting threads, approx. 2 meters long
- 1 tall rod (or pair of ceiling hooks, for suspending the ball)
- 1 prod (metal-coated ping pong ball with an insulating handle)
- 1 electronic balance (for finding the mass of suspended sphere)
- 1 ruler or meter stick

You will also need computer-based video capture and video analysis capability:

- 1 computer
- 1 digital video camera
- 1 video capture software
- 1 video analysis software

Recommended group size:	4	Interactive demo OK?:	Y

Note: If you are using a computer-based video analysis system, you must use one of the video workstations and special video capture software to make your movie. Your instructor will provide you with the information to do video capture.

Setting up the Force Law Experiment

The purpose of this experiment is to verify that the magnitude of the electrical interaction forces between two small metal-coated balls varies as $1/r^2$, where r is the distance between the balls. To obtain data using a digital videocamera attached to a computer video capture card:

1. Suspend the metal-coated ping pong ball from two long (about 2 meters) nylon threads. Record the vertical distance from the point of suspension to the center of the hanging sphere. Place a meter stick horizontally under the hanging ball.
2. Use fur to charge a plastic rod and transfer charge from the rod to both the prod and the hanging ball.
3. *Carefully touch the ball with the prod so that the two balls have the same amount of charge.*
4. Practice bringing the charged prod closer and closer to the hanging ball, slowly and steadily. The trick is to *keep the line between the ball and the prod horizontal at all times.*
5. Once you get good at step 4, repeat it two or more times while the video camera is running. Start each movie with the prod very far away from the ball so that there is no noticeable interaction between the two.

Creating Digital Video Frames with a Computer-based Video Capture System

1. Capture and save about 15 frames covering distances between the ball and prod from "infinity" (where no noticeable interaction) to about 2 cm. **Note:** You probably have about 100 or so frames in each segment—there is no need to save them all!
2. Refer to the instructions for the video capture software to configure it for transforming your digital video frames to the QuickTime™ format and compressing the frames.

Analyzing the Digital Images Video to Find Distances and Forces

1. Pick your best sequence of digital video frames for analysis.
2. Use the video analysis software to determine: (i) the distance, r, between the charged objects and (ii) the distance, x, from the suspended ball to a vertical line (to determine the angle θ).
3. The distances and/or coordinates you measure will be in pixels*. If you have video analysis software with scaling capability, you should use it. If not, you will need to find the scaling factors you must multiply the pixels by to get real laboratory units (cm or m). To do this, determine the number of pixels between the end points of an object of known length (such as the length of a meter stick placed in the field of view) in each direction in the video frame.
4. Determine the effective length of the strings suspending the ball (that is, the vertical distance from the line of suspension to the hanging ball).

*The term pixel stands for *picture element* and is the fundamental unit used to locate points on the screen of a computer monitor.

5. Use your video analysis software to determine the horizontal distance between the prod and the suspended ball for each frame.
6. Either export your frame-by-frame video data to a spreadsheet file for analysis or use tools in your video analysis software to complete an analysis.

19.6.2. Activity: Verifying Coulomb's Law Experimentally

1. Use the equation you derived in Activity 19.6.1.f to calculate the experimental value of the electrostatic force in each frame from your data.
2. Analyze your data using a video analysis or a spreadsheet model and present your results in tabular and graphical form. In particular, try modeling your plot of F^{elec} vs. r with an equation of the form $F^{elec} = C/r^2$ where C is a constant.
3. Draw conclusions. Does the $F^{elec} = C/r^2$ relationship seem to hold?
4. Describe the most plausible sources of uncertainty in your data.

19.6.3. Activity: How Much Charge Is on the Hanging Sphere?

Since you touched the ball and prod together before starting, they should each have the same amount of charge. In your experiment, a model of F vs. r should have the form

$$F^{elec} = \frac{k|q_A||q_B|}{r^2} = kq^2 \frac{1}{r^2}$$

where in this special case $q = q_A = q_B$. We use absolute value signs to signify the amount of charge regardless of whether it is positive or negative. You should be able to calculate the amount of charge on the ball (and probe). Do so!

THE ELECTRIC FIELD

19.7. THE ELECTRIC FIELD

Until this week, most of the forces you studied resulted from the direct action or contact of one piece of matter with another. From your direct observations of charged, metal-coated balls, it should be obvious that charged objects can exert electrical forces on each other at a distance. How can this be? The action at a distance that characterizes electrical forces, or for that matter gravitational forces, is in some ways inconceivable to us. How can one charged object feel the presence of another and detect its motion with only empty space in between? Since all atoms and molecules are thought to contain electrical charges, physicists currently believe that all "contact" forces are actually electrical forces involving small separations. So, even though forces acting at a distance seem inconceivable to most people, physicists believe that all forces act at a distance.

Fig. 19.11.

Physicists now explain all forces between charged particles, including contact forces, in terms of the transmission of traveling electromagnetic waves. We will engage in a preliminary consideration of the electromagnetic wave theory toward the end of the semester. For the present, let's consider the attempts of Michael Faraday and others to explain action-at-a-distance forces back in the nineteenth century. Understanding more about these attempts should help you develop some useful models to describe the forces between charged objects in some situations.

To describe action at a distance, Michael Faraday introduced the notion of an *electric field* emanating from a collection of charged objects and extending out into space. More formally, the electric field due to a known collection of charged source objects is represented by an electric field vector at every point in space. Thus, the electric field vector due to the collection of source charges, $\vec{E}_s$, is defined as the force, $\vec{F}_{s\to t}$, experienced by a very small positive test charge at a point in space divided by the magnitude of the test charge q_t. The electric field is in the direction of the force e on a small positive "test" charge and has the magnitude of

$$\vec{E}_s \equiv \frac{\vec{F}_{s\to t}}{q_t}$$

where q_t is the charge on a small test particle and $\vec{F}_{s\to t}$ is the net force on the test charge due to the source charges. In general, the $\vec{E}_s$ vector will have a different magnitude and direction for each possible location of the test charge.

To investigate the vector nature of an electric field, you can use a positively charged, metal-coated ball, suspended from a string, as the test charge. (The ball is charged by touching it with a glass rod that has been rubbed with polyester.) Charge up the glass rod and hold it in a vertical position. The charge on the glass rod is the source of the electric field. Now hold the test charge by its string and move it around the rod. Note the direction and magnitude of the force at various locations around the rod. What is the direction and relative magnitude of the electric field around the rod? To complete the suggested observations you will need the following:

- 1 threaded, metal-coated Styrofoam ball (with low mass)
- 1 plastic rod
- 1 fur
- 1 glass rod
- 1 polyester cloth
- 1 ruler

Recommended group size:	2	Interactive demo OK?:	N

Note: By convention physicists always place the tail of the E-field vector at the point in space of interest rather than at the charged object that causes the field.

19.7.1. Activity: Electric Field Vectors from a Positively Charged Rod

Make a qualitative sketch of some electric field vectors around the rod at the points in space marked on the following diagram. The length of each vector should roughly indicate the *relative* magnitude of the field (that is, if the E-field is stronger at one point than another, make its vector longer). Of course, the direction of the vector should indicate the direction of the field. Don't forget to put the tail of the vector at the location of interest, not at the location of the glass rod.

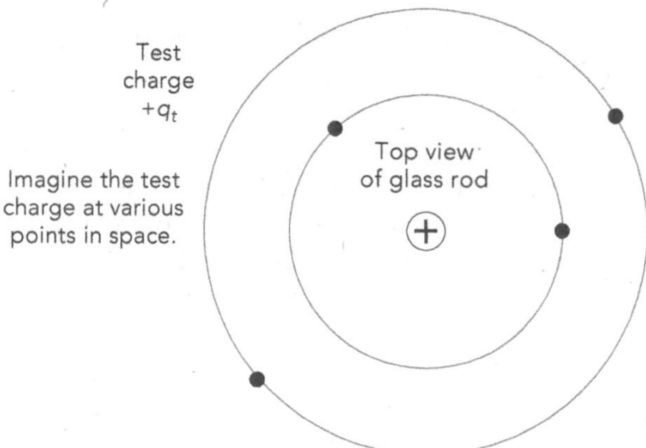

19.7.2. Activity: Electric Field from a Negatively Charged Rod

Use the hard plastic rod to create an electric field resulting from a negative charge distribution. Sketch the electric field vectors at the indicated points in space; show both the magnitude and direction of the vectors.

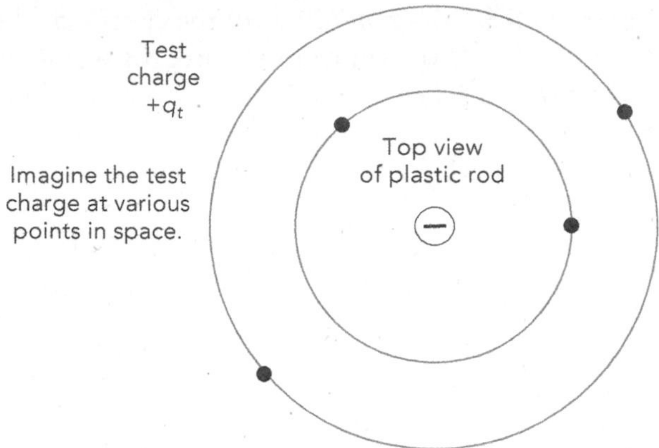

19.8. SUPERPOSITION OF ELECTRIC FIELD VECTORS

The fact that electric fields from charged objects that are distributed at different locations act along a line between the charged objects and the point in space of interest is known as *linearity*. The fact that the vector fields due to charged objects at different points in space can be added together vectorially is known as *superposition*. These two properties of the Coulomb force and the electric field that derives from it are very useful in our endeavor to calculate the value of the electric fields due to a collection of point charges at different locations. This can be done by finding the value of the E-field vector from each point charge and then using the *principle of superposition* to determine the vector sum of individual electric field vectors created by each source charge in a collection.

19.8.1. Activity: Electric Field Vectors from Two Point Charges

a. Look up the equations for Coulomb's law and the electric field from a point charge in Section 19.4. Also check out the value of the Coulomb constant you need to calculate the actual value of the electric field from a point charge. List the equations and the Coulomb constant in the space below.

b. Use a spreadsheet to calculate the *magnitude* of the electric field (in N/C) at distances of 0.5, 1.0, 1.5 . . . 10.0 cm from a source consisting of a single point-like object with an excess charge of 2.0×10^{-9} C. Be careful to use the correct units (that is, convert the distances to meters before doing the calculation). Affix the results below for later reference.

c. The following graph shows an *E*-field source consisting of two point-like particles with excess charges of $+2.0 \times 10^{-9}$ C and -2.0×10^{-9} C respectively that are separated by a distance of 8.0 cm. Draw the vector contribution of each of the point charges to the electric field at each of the four points shown below where a test charge q_t could be located in space. Use your spreadsheet results and a scale in which the vector is 1 cm long for each electric field magnitude of 1.0×10^4 N/C. Then use the principle of superposition and the polygon method to find the resultant $\vec{E}_s$ vector at each point. **Hint:** One of the source charges will attract a positive test charge and the other will repel it.

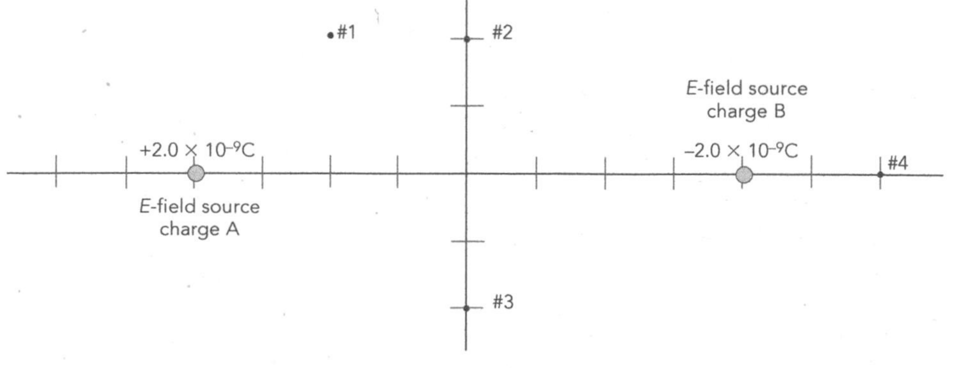

19.9. THE ELECTRIC FIELD FROM AN EXTENDED CHARGE DISTRIBUTION

Although electric charges will not usually distribute themselves uniformly throughout a conductor, charge can be distributed uniformly through an insulator. If a total excess charge of q^{tot} is distributed uniformly throughout a continuous extended insulated object, it can be divided into small segments each of which contains a charge q. Then, by assuming that each segment behaves like a tiny point-like charge, the electric field at a point P in space due to each segment can be calculated. The total electric field at P is simply the vector sum of the contributions of each of the charge segments. This process yields an approximate value of the electric field at point P. Such approximate values can be calculated quite readily using a computer spreadsheet. To get a more exact value we must sum up infinitely many infinitesimally small elements of charge Δq. This is what mathematical integration is all about.

The goal of this section of the activity guide is to calculate the electric field $\vec{E}$ corresponding to a continuous charge distribution on a rod at two points in space, P and P', as shown below.

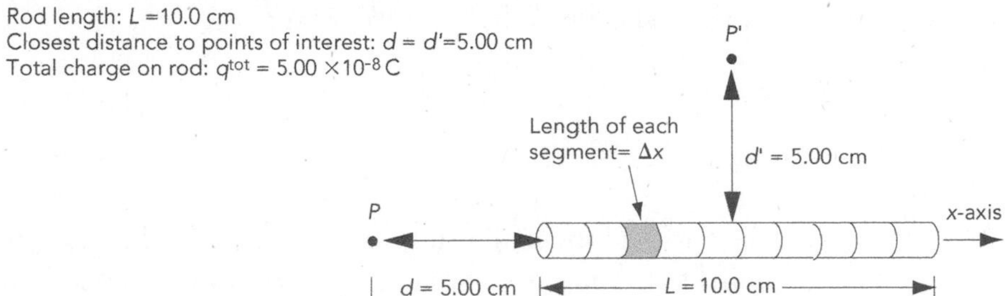

Fig. 19.12. Diagram of a charged rod broken into ten imaginary "point" charges for a spreadsheet calculation of the electric field at points P and P'.

Each of these calculations will be done two ways: (1) doing an approximate numerical calculation with the spreadsheet, and (2) doing an "exact" integration. These two methods of calculation will be compared with each other. You could extend the calculation to other points in space and graph the change in field as a function of the distance from the rod along a line through the axis of the rod and along a line perpendicular to the rod.

19.9.1. Activity: E-Field Vectors from a Uniformly Charged Rod

In each case, draw the magnitude and direction of ten vectors, $\vec{E}_n$ which represent $\vec{E}_1$, $\vec{E}_2$, $\vec{E}_3$, etc. at point P or P'. Each vector approximates the relative scale of each $\Delta\vec{E}_n$ at the two points (draw longer arrows for the vectors corresponding to charge elements closer to P or P'). Use the following diagrams and draw a resultant vector in each case. **Note:** It is customary to locate the tail of each E-field vector at point P where our imaginary test charge q_t is located (rather than at the location of each source charge).

a. Parallel to the axis of the rod

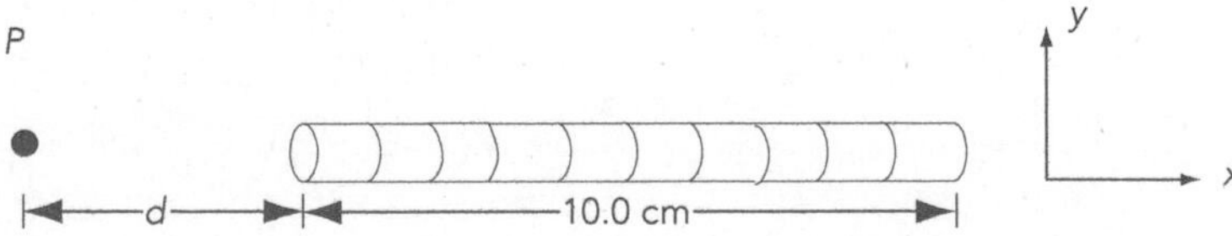

b. Perpendicular to the axis of the rod

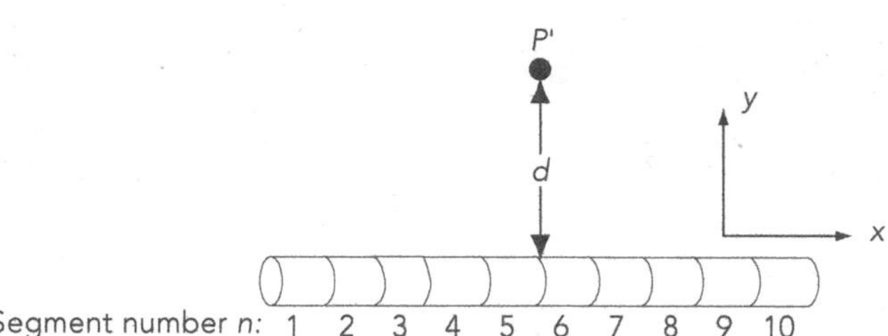

19.9.2. Activity: Electric Field Calculations Along the Axis of a Rod

a. Consider a rod of length L that is divided into n segments. If the total charge on the rod is given by q^{tot}, show that the charge $q = q_i$ in each segment Δx of the rod is given by $q = (q^{tot}/L)\,\Delta x$.

b. Use a spreadsheet to find the magnitude of the electric field at the point P which is parallel to the axis of the rod. The contribution of each of the ten rod elements to the electric field at point P can be found numerically by using distance and length data for each element (see Figure 19.12.). You should set up three columns: Charge Segment number, Δx, and $\vec{E}_n$. Once the calculations are done you can sum up the ten $\vec{E}_n$ values vectorially to get the value of $\vec{E}^{tot}$. Affix a printout of your spreadsheet in the space below. **Hints:** (1) The Coulomb constant $k = 9.0 \times 10^9$ N·m^2/C^2 and (2) in this simple 1D situation $\vec{E}_n = E_{n\,x}\hat{\imath}$ so there is no y-component of the electric field at point P. (3) $\vec{E}^{tot} = \sum_{n=1}^{10} \vec{E}_n = (E_{1\,x} + E_{2\,x} + \ldots E_{10\,x})\hat{\imath}$ so the x-component of the total E-field at point P is $E_x^{tot} = E_{1\,x} + E_{2\,x} + \ldots E_{10\,x}$.

Deriving an "Exact" Equation for a Point Along the Rod Axis

The calculation you just completed provided a numerical approximation to the electric field on the axis of the rod for a specific situation. It is possible to use integration to derive a general equation describing magnitude of the electric field at any point P along the axis of the rod. In particular, you can show by integration that the magnitude of the electric field at a distance d from the end of a rod of length L is given by

$$E = \frac{kq^{\text{tot}}}{d(L+d)} \tag{19.3}$$

where k is the Coulomb constant and q^{tot} is the total charge on the rod. You will begin by dividing the rod into infinitely many elements of infinitesimal length dx. Next you will derive an equation for the electric field contribution of each element. Finally, you will set up and solve an integral that represents the sum of all these elements.

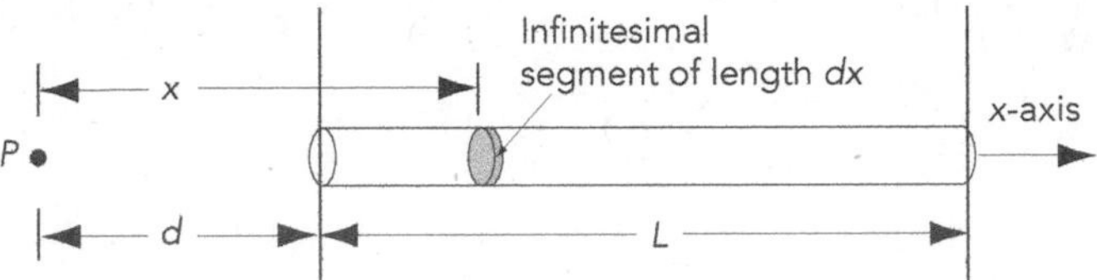

Fig. 19.13. Diagram showing an infinitesimal length dx at an arbitrary distance along the rod of length L. The distance from point P to the segment is denoted as x.

19.9.3. Activity: Using Integration to Find *E* Along the Axis of a Rod.

a. Explain why the fraction of charge, q, on each infinitesimal length dx on the rod is given by $dq = q^{\text{tot}}(dx/L)$ where q^{tot} is the total charge on the rod.

b. Refer to Figure 19.13. to explain why the x-component of the electric field, dE_x, at point P due to a rod segment of infinitesimal width dx that is a distance x from point P is given by the equation $dE_x = (kq^{\text{tot}}/x^2L)dx$.

c. An integral presents the sum of many infinitely small elements. Thus the x-component of the electric field E at point P can be represented by the integral

$$E_x^{\text{tot}} = \int_{x^{\text{min}}}^{x^{\text{max}}} \frac{kq^{\text{tot}}}{x^2L}\,dx$$

Explain why $x^{\text{min}} = d$ and $x^{\text{max}} = d + L$.

d. Explain why k, q^{tot}, and L can be pulled out of the integral so that

$$E_x^{tot} = \frac{kq^{tot}}{L}\int_d^{d+L} \frac{1}{x^2}\,dx.$$

e. Perform the integration to show that $E_x^{tot} = \frac{kq^{tot}}{L}\int_d^{d+L} \frac{1}{x^2}dx = \frac{kq^{tot}}{d(L+d)}$.

f. Calculate the "exact" value for the electric field magnitude E at point P by substituting the values for L, d, and q^{tot} shown in Fig. 19.12. into the equation

$$E_x^{tot} = \frac{kq^{tot}}{d(L+d)} =$$

g. How does the numerical value you calculated in 19.9.1.b. compare to the "exact" value you just calculated? Compute the percent discrepancy. How could you make the numerical method more "exact?"

The first set of calculations for the E-field along the axis of the rod was relatively easy because all the electric field vectors lie along a single line. In order to do a calculation of the E-field at point P' perpendicular to the axis of the rod that lies along a line bisecting the rod, we have to consider both the x- and y-components of the electric field resulting from the charge on each element. Thus, in general, $\vec{E}_n = E_{n\,x}\hat{\imath} + E_{n\,y}\hat{\jmath}$.

19.9.4. Activity: The Electric Field $\perp$ to the Axis of the Rod

Explain why the x-component of the total E-vector at point P' should be zero. **Hint:** The argument used is known as a symmetry argument.

Name ______________________ Section ____________ Date ____________

UNIT 20: ELECTRIC FLUX AND GAUSS' LAW

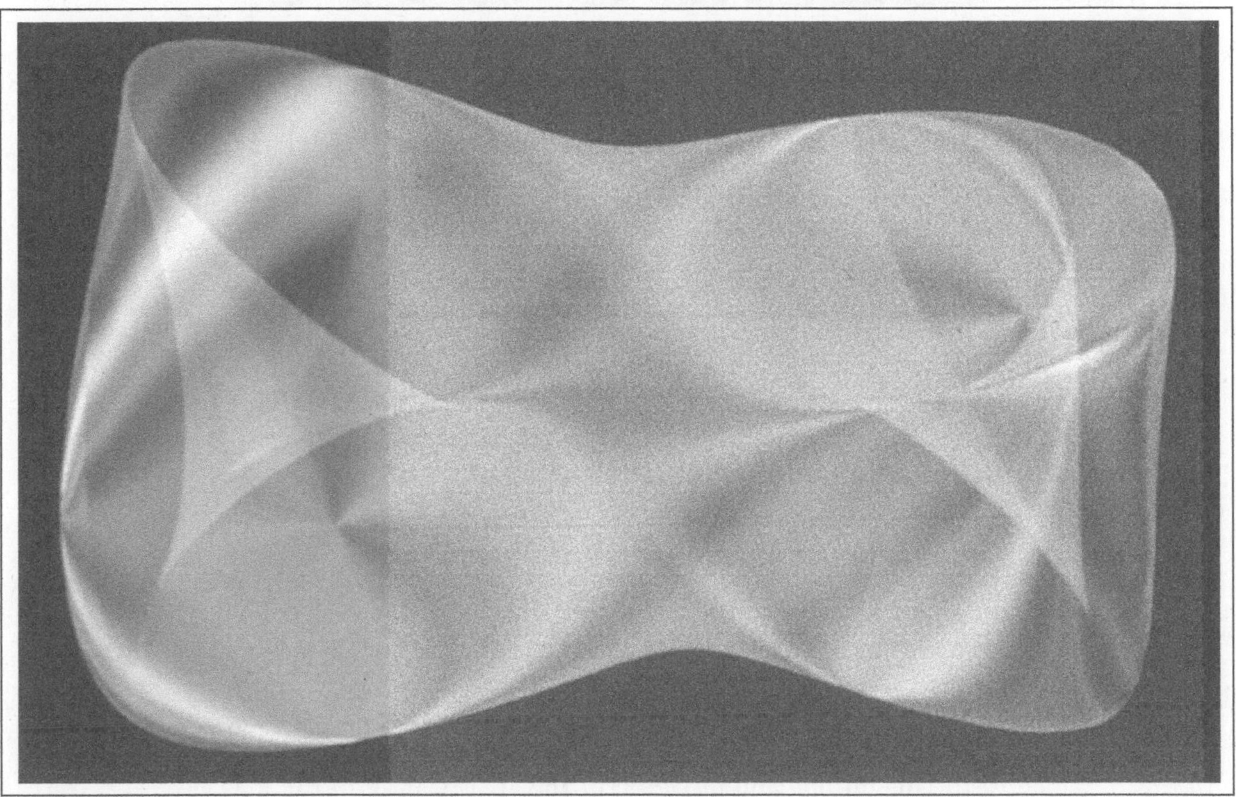

The complex figure shown above has a closed surface that contains a volume. If we place any number of positive and negative charges in the volume contained by an imaginary complex surface of known area, we can use Gauss' law to find the average value of the components of electric field perpendicular to the surface. This law is of great academic interest because it is based on the mathematical nature of electrical forces. However, when charges are placed in certain simple configurations, the application of Gauss' law can provide a very powerful way to find the value of the net electric field anywhere in space resulting from those charges. Upon completing this unit, you should appreciate how useful a seemingly arcane abstraction can be.

UNIT 20: ELECTRIC FLUX AND GAUSS' LAW

... before Maxwell people considered physical reality. . . . as material points. . . . After Maxwell they considered physical reality as continuous fields. . . .
Albert Einstein

OBJECTIVES

1. To understand how electric field lines and electric flux can be used to describe the magnitude and direction of the electric field in a small region in space.

2. To discover how the electric flux passing through a small area is related to the magnitude and direction of the area relative to the magnitude and direction of the electric field lines.

3. To discover the relationship between the flux passing through a "closed surface" and the charge enclosed by that surface for a two-dimensional situation (Gauss' law in Flatland).

4. To review how the expression $\int \vec{E} \cdot d\vec{A}$ over a closed three-dimensional surface is proportional to the number of field lines passing through the closed surface and thus to extend the discovery of Gauss' law in Flatland to three dimensions.

5. To explore the concept of symmetry.

6. To learn to use Gauss' law to calculate the electric fields that result from highly symmetric distributions of electric charge at various points in space.

20.1. OVERVIEW

Coulomb's law and the principle of superposition can be used to calculate the force, and hence the electric field, on a test charge due to charge distributions that surround it. It is possible, however, to calculate the electric field using a completely different formulation of Coulomb's law. This formulation is known as Gauss' law and it involves relating the field surrounding a collection of charges to the amount of charge enclosed by a surface. The Gauss' law formulation is a very powerful tool for calculating electric fields due to *symmetric* distributions of charge. By using the rules of integration and vector algebra, Gauss' law can be proven to be mathematically equivalent to Coulomb's law.

You will begin the study of Gauss' law by learning about a convenient construct known as *electric field lines*. These lines can be used to map the direction of the net force on a small test charge at any point in space due to other charges. You will practice constructing the electric field lines from a configuration of charges using superposition and Coulomb's law.

Next you are going to take a non-mathematical approach to discovering a two-dimensional Gauss' law. You will use a computer simulation to create a pattern of electric field lines around a collection of charges that you create. You will then draw closed "surfaces" around various charges or groups of charges and see how many electric field lines pass in and out of the surfaces. Finally, you can explore the concept of symmetry and use the mathematical representation of Gauss' law in three dimensions to calculate the electric field at various points in space due to uniform charge distributions.

ELECTRIC FIELD LINES AND FLUX

20.2. ELECTRIC FIELD LINES

You have been representing the electric field due to a configuration of electric charges by an arrow that indicates magnitude and direction; using the principles of superposition and linearity, you can determine the length and direction of the arrow for each point in space. This is the *conventional representation* of a "vector field." An alternative representation of the vector field involves defining *electric field lines.* Unlike an electric field vector, which is an arrow with magnitude and direction, electric field lines are continuous. You can use a simulation written for the Macintosh computer by Blas Cabrera at Stanford University to explore some of the properties of electric field lines for some simple situations.

For this activity you will need:

- 1 computer
- 1 Coulomb computer simulation

Recommended group size:	2	Interactive demo OK?:	Y

20.2.1. Activity: Simulation of Electric Field Lines from Point Charges

Load the Coulomb program into the computer. Figure out how to turn off the axes and then run the simulations outlined below with the axes off.

a. Set a single charge of ±1, 2, 3, or 4 Coulombs somewhere on the computer screen and run the program. After a few minutes sketch the lines in the space below. Show the direction of the electric field on each line by placing arrows on them. Indicate how much charge you used on the diagram.

How many lines are there in the drawing? Are the lines more dense or less dense near the charge? Explain. How does the direction of the lines depend on the sign of the charge?

b. Try another magnitude of charge. No need to sketch the result, but how many lines are there? Can you describe the rule for telling how many lines will come out of, or start or end on, a charge in this simulation if you know the magnitude of the charge in Coulombs?

c. Repeat the exercise using two charges with the same magnitude having unlike signs. Place them at two different locations on the screen. After a few minutes sketch the lines in the space below. Indicate how much charge you used on the diagram. Comment on why the lines are more or less dense near the charges. How does the direction of the lines depend on the sign of the charge?

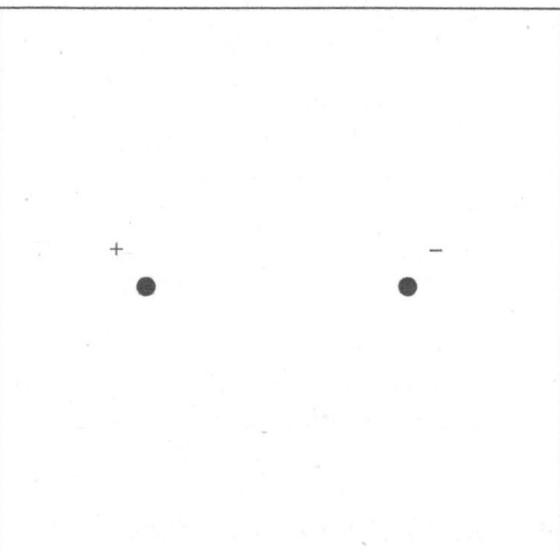

d. Summarize the properties of electric field lines. What does the number of lines signify? What does the direction of a line at each point in space represent? What does the density of the lines reveal?

20.3. ELECTRIC FLUX

As you have just seen, we can think of an electrical charge as having a number of electric field lines, either converging on it or diverging from it, that is proportional to the magnitude of its charge. We can now explore the mathematics of enclosing charges with surfaces and seeing how many electric field lines pass through a given surface. *Electric flux* is defined as a measure of the number of electric field lines passing through a surface. In defining "flux" we are constructing a mental model of lines streaming out from the surface area surrounding each unit of charge like streams of water or rays of light. Modern physicists do not really think of charges as having anything real streaming out from them, but the mathematics that best describes the forces between charges is the same as the mathematics that describes streams of water or rays of light. So, for now, let's explore the behavior of this model.

It should be obvious that the number of field lines passing through a surface depends on how that surface is oriented relative to the lines. The orientation of a small surface of area A is usually defined as a vector that is perpendicular to the surface and has a magnitude equal to the surface area. By convention, the normal vector points away from the *outside* of the surface. The normal vector is pictured below for two small surfaces of area A and A' respectively. In the picture it is assumed that the outside of each surface is white and the inside grey.

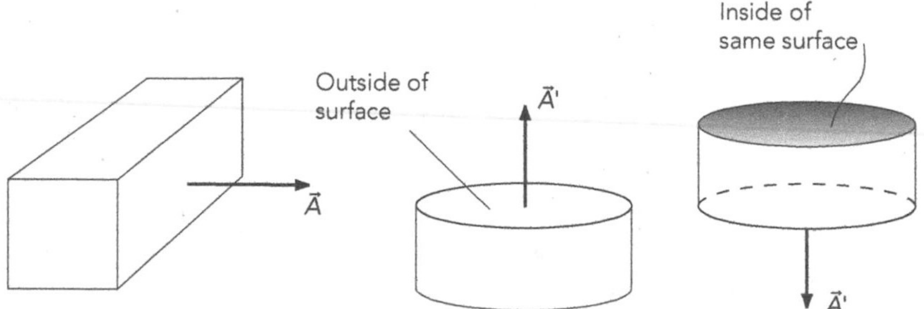

Fig. 20.1. A diagram showing how the normal vector representing an area points from inside to outside. The inside of a surface is shaded grey.

20.3.1. Activity: Drawing Normal Vectors

Use the definition of "normal to an area" given above to draw normals to the surfaces shown below. Let the length of the normal vector in cm be *equal in magnitude to each area in cm²*. Don't just draw arrows of arbitrary length.

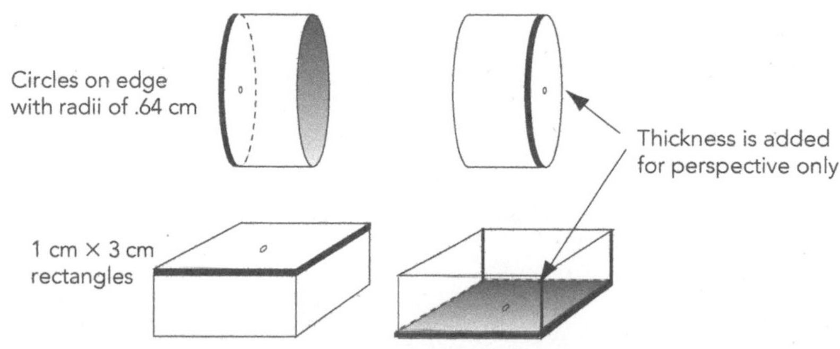

Fig. 20.2.

By convention, if an electric field line passes from the inside to the outside of a surface, we say the flux is positive. If the field line passes from the outside to the inside of a surface, the flux is negative.

How does the flux through a surface depend on the angle between the normal vector to the surface and the electric field lines? In order to answer this question in a concrete way, you can make a mechanical model of some electric field lines and of a surface. What happens to the electric flux as you rotate the surface at various angles between 0 degrees (or 0 radians) and 180 degrees (or π radians) with respect to the electric field vectors? To make your model you will need to arrange nails in a 10 × 10 array poking up at 1/4" intervals through a piece of Styrofoam. The surface can be a copper loop painted white on the "outside." You will need:

- Styrofoam or 3/8" plywood (5" × 5" square)
- 100 nails, approximately 4" in length (mounted on the Styrofoam or plywood)
- 1 wire loop (4" × 4" square)
- 1 paper, 5" × 5" with 1/4" graph rulings (to affix to the mounting square to help with spacing the nails)
- 1 protractor

Recommended group size:	4	Interactive demo OK?:	N

Once the model is made you can perform the measurements with a protractor and enter the angle in radians and the flux into a computer data table for graphical analysis.

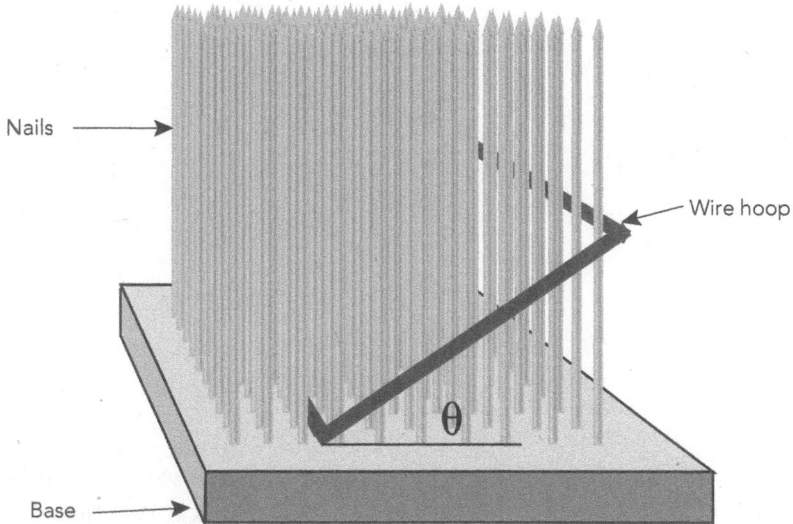

Fig. 20.3. Apparatus designed to determine how many uniformly space flux lines will pass through an imaginary surface area as a function of the angle between the direction of the flux lines and the normal vector representing the surface area.

20.3.2. Activity: Flux as a Function of Surface Angle

a. Use your mechanical model, a protractor, and some calculations to fill in the following data table, or you may want to feed data directly into your graphing software for later display. **Hint:** Interested in being lazy but clever? Use symmetry to determine the angles corresponding to negative flux without making more measurements.

θ Degrees	θ Radians	Φ Number of lines
		100
		90
		80
		70
		60
		50
		40
		30
		20
		10
		0
		-10
		-20
		-30
		-40
		-50
		-60
		-70
		-80
		-90
		-100

b. Plot the flux Φ as a function of radians. Look *very* carefully at the data. Is it a line or a curve? What mathematical function might describe the relationship?

c. Try to confirm your guess by constructing a spreadsheet model and overlay graph of the data and the mathematical function you think matches the data. Affix or sketch the plots in the space to the right of the data table. Summarize your procedures and conclusions in the following space.

d. What is the definition of the vector dot product of two vectors in terms of vector magnitudes and the angle θ between them? Can you relate the scalar value of the flux, Φ, to the dot product of the vectors $\vec{E}$ and $\vec{A}$?

20.4. A MATHEMATICAL REPRESENTATION OF FLUX THROUGH A SURFACE

One convenient way to express the relationship between angle and flux for a uniform electric field $\vec{E}$ is to use the dot product so that the flux through a surface $\vec{A}$ is $\Phi = \vec{E} \cdot \vec{A}$. Flux is a scalar. If the electric field is not uniform or if the surface subtends different angles with respect to the electric field lines, then we must calculate the net or total flux by breaking the surface into infinitely many infinitesimal areas, $d\vec{A}$, so that $d\Phi = \vec{E} \cdot d\vec{A}$, and then taking the integral to sum all the flux elements. This gives a net flux of

$$\Phi^{\text{net}} = \int d\Phi = \int \vec{E} \cdot d\vec{A} \quad \text{(net flux through a surface)}$$

Some surfaces, like that of a sphere or the series of surfaces that make a rectangular box, are closed surfaces. A *closed surface* has no holes or edges so that nothing can leave its interior without passing through the surface itself. Because we want to study the amount of flux passing through closed surfaces, there is a special notation to represent the integral of $\vec{E} \cdot d\vec{A}$ through a closed surface. It is represented as follows:

$$\Phi = \oint d\Phi = \oint \vec{E} \cdot d\vec{A} \qquad \text{(net flux through a closed surface)}$$

GAUSS' LAW

20.5. DISCOVERING GAUSS' LAW IN FLATLAND

How is the flux passing through a closed surface related to the enclosed charge? Let's pretend we live in a two-dimensional world in which all charges and electric field lines are constrained to lie in a flat two-dimensional space—of course, mathematicians call such a space a plane.[1]

For this project you will need:

- 1 computer
- 1 Coulomb software simulation[2]

Recommended group size:	2	Interactive demo OK?:	N

Open the Coulomb program on the computer again and set it to sketch lines for some nutty, creative mix of charges. Don't be *too* creative or the lines will take forever to sketch out. You should do the following:

1. Open the Coulomb simulation and place some positive *and* negative charges at different places on the screen. Then start the program to calculate and display the electric field lines in two dimensions.
2. Either sketch or print out the screen configuration showing the charges and the associated "E-field" lines.
3. Draw arrows on each of the lines indicating in what direction a *small* positive test charge would move. **Note:** "Small" means that the test charge does not exert enough forces on the charge distributions that create the E-field to cause the field to change noticeably.
4. Figure out what the two-dimensional equivalent of a "closed surface" ought to look like and draw several "closed surfaces" on your diagram.

Fig. 20.4.

[1] If you haven't already read it, we recommend that you read E. A. Abbot's book entitled *Flatland; A Romance of Many Dimensions* (Dover, New York, 1952). It's a delightful piece of late nineteenth-century political satire in the guise of a mathematical spoof.

[2] In a two-dimensional map of flux lines it is not possible to assign a fixed number of lines to a charge and to assign spacings between coming from infinity unambiguously. Thus, the line densities may not look proper in some of the Coulomb software plots. This should not matter to students completing this exercise.

Having done all of this preparation, you should be ready to discover how the net number of lines passing through a surface is related to the net charge enclosed by the surface.

20.5.1. Activity: Gauss' Law in Flatland

a. Place a replica of the charge configuration you designed and the associated field lines in the space below.

b. Draw some two-dimensional closed "surfaces" in pencil in the space above. Some of them should enclose charge, and some should avoid enclosing charge. Count net flux lines coming out of each "surface." **Note:** Consider lines coming out of a surface as positive and lines going into a surface as negative. The *net number of lines* is defined as the number of positive lines minus the number of negative lines.

	Charge enclosed by the arbitrary surface			Lines of flux in and out of the surface		
	Total Positive Charge (Arbitrary Units)	Total Negative Charge (Arbitrary Units)	q^{net}	Φ_{out}	Φ_{in}	Φ^{net}
1						
2						
3						

c. What is the apparent relationship between the net flux and the net charge enclosed by a two-dimensional "surface"?

Gauss' Law in Three Dimensions

If you were to repeat the simulated exploration you just performed in a three-dimensional space, what do you think would be the appropriate expression for Gauss' law?

20.5.2. Activity: Statement of Gauss' Law

a. Express the three-dimensional form of Gauss' law in words.

b. Express the law using an equation.

20.6. ELECTRIC FIELDS AND CHARGES INSIDE A CONDUCTOR

An electrical conductor is a material that has electrical charges in it that are free to move. If a charge in a conductor experiences an electric field, it will move under the influence of that field since it is not bound (as it would be in an insulator). Thus, we can conclude that if there are no moving charges inside a conductor, the electric field in the conductor must be zero.

Let's consider a conductor that has been touched by a charged plastic rod so that it has an excess of negative charge on it. Where does this charge go if it is free to move? Is it distributed uniformly throughout the conductor? If we know that $\vec{E} = 0$ inside a conductor, we can use Gauss' law to figure out where the excess charge on the conductor is located.

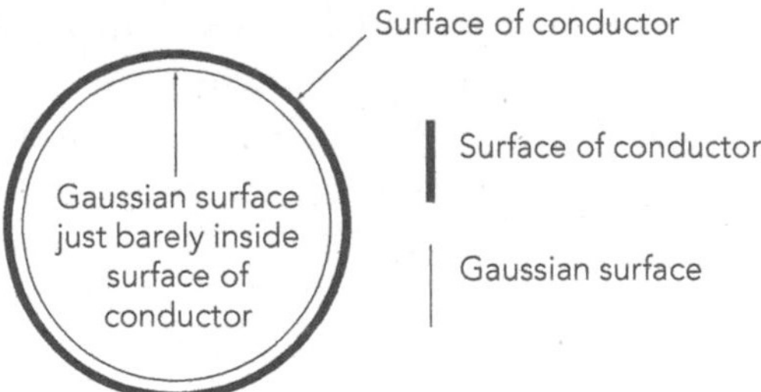

Fig. 20.5. Diagram showing how an imaginary Gaussian surface can be constructed just inside the surface of the conductor.

20.6.1. Activity: Where Is the Excess Charge in a Metal?

a. Consider a conductor with an excess charge of Q. If there is no electric field inside the conductor (by definition), then what is the amount of excess charge enclosed by the Gaussian surface just inside the surface of the conductor? **Hint:** Use the three-dimensional Gauss' law here.

b. If the conductor has excess charge and it can't be inside the Gaussian surface according to Gauss' law, then what's the only place the charge can be?

c. Given the fact that as like charges, the excess charges will repel each other, is the conclusion you reached in part b. above physically reasonable? Explain. **Hint:** How can each unit of excess charge that is repelling every other unit of excess charge get as far away as possible from the other excess charges on the conductor?

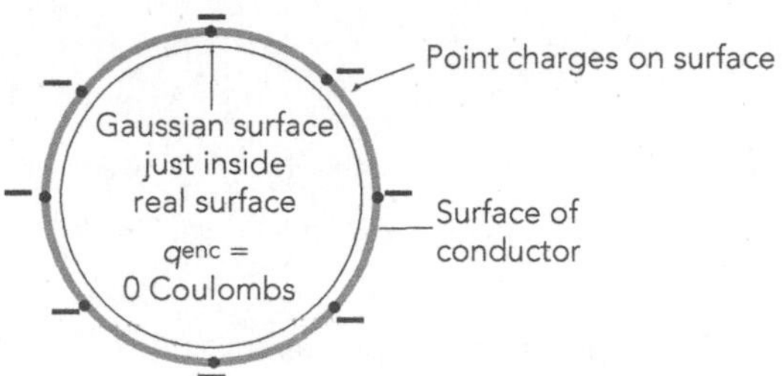

Fig. 20.6.

20.7. EXPERIMENTAL CONFIRMATION OF GAUSS' LAW

You used Gauss' law to predict that excess charge in a conductor would move to the outside surface of the conductor. Let's check this prediction. We can do this using:

- 1 soup can (a surrogate ice pail)
- 1 black plastic rod
- 1 fur
- 1 metal-coated, threaded ball (with low mass)
- 1 electroscope with gold leaves (optional)

Recommended group size:	4	Interactive demo OK?:	N

The test of Gauss' law that you are about to perform is attributed to Benjamin Franklin but is usually referred to as the Faraday Ice Pail Experiment. How did Faraday get credit for this one?

20.7.1. Activity: The Faraday Ice Pail Experiment

a. Suppose that a plastic rod that has been rubbed with fur is used to charge a metal-coated ball. Mark the sign of the charges on the ball in the following diagram.

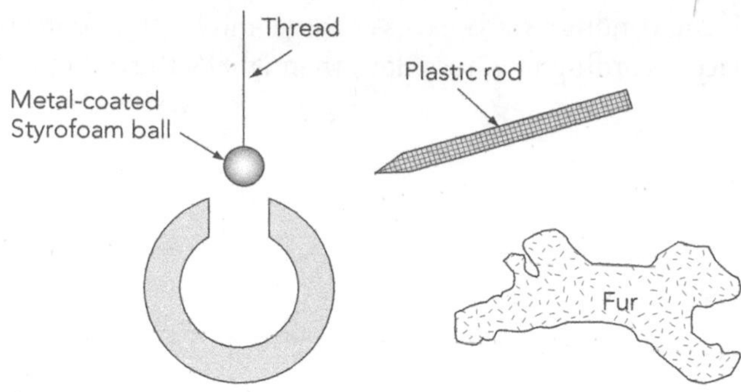

Spherical metal can with hole

b. The charged ball is lowered into a metal can. If negative charges are free to move inside the can leaving positive charges unneutralized, what kind of excess charge will be on the inside of the can? What will be on the outside? Mark the sign of the charge on the ball and on the inside and the outside of the can.

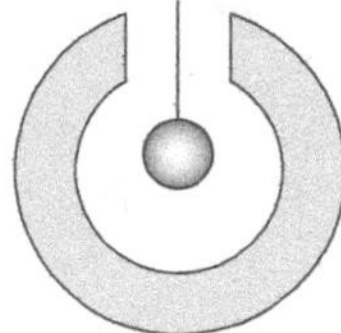

c. What will happen to the charges on the ball and the inside of the can if the ball touches the inside of the can? How about the charges on the outside of the can? Draw the location of the charges.

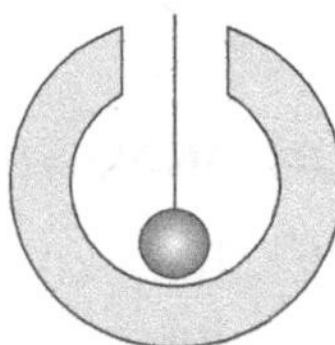

d. After the ball is removed, is there an E-field in the can? Outside of the can? **Hint:** Is charge enclosed by an inner Gaussian surface? An outer one?

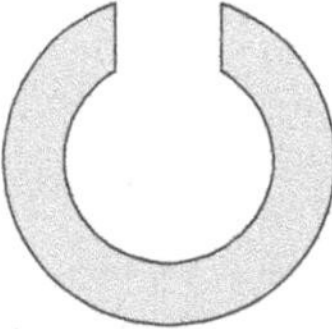

e. Suppose that the ball is charged up again and brought near the can. Will it be attracted to or repelled from the outside of the can? Why or why not? Will it be attracted to the inside of the can? Why or why not?

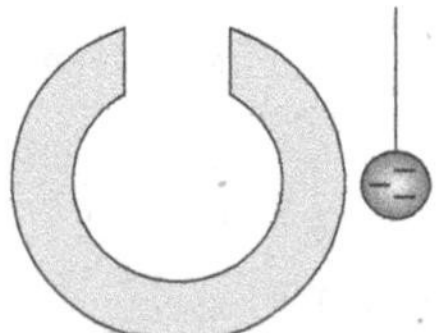

f. If you have an electroscope available, check to see if the outside of the can is actually charged.

g. Did your observations match your prediction in part e?

h. Explain why Gauss' law provides a basis for explaining your observations.

A Gauss' Law Puzzler

The Faraday ice pail activity provides a very powerful demonstration of the application of Gauss' law to a real situation. The following puzzler should give you some more practice with applying this law.

20.7.2. Activity: A Faraday Ice Pail Puzzler

a. Can you devise a way to use the same negatively charged rod as you used in the last activity to charge the can positively? Talk this one over with your partners and explain your method.

b. Devise a way to test the validity of your method. Explain your test. Did your method pass the test?

GAUSS'S LAW CALCULATIONS

20.8. SYMMETRY

We're making all this fuss about Gauss' law so that you can calculate the electric field from *symmetric* charge distributions. What is a "symmetric" distribution? It's an arrangement of charges that can be rotated about an axis and/or reflected in a mirror and still look the same. Let's pretend that the objects below are made of rigid insulating materials and that charge is uniformly distributed throughout each object. Based on the *external* appearance of the objects, which ones appear to have symmetry? Which objects don't appear symmetric? Don't worry about the internal structure of the objects. Some of these situations are subtle. For example, does the mirror image of a stopwatch turn the same way?

20.8.1. Activity: Symmetric Charge Distributions

Suppose charges were distributed evenly throughout the objects in Figure 20.7. Circle the symmetric charge distributions. Don't forget to look at the letters!

A B C D E F G H I J K L M N

Fig. 20.7.

20.9. USING GAUSS' LAW TO CALCULATE ELECTRIC FIELDS

Gauss' law can be stated mathematically by means of the expression

$$\Phi^{net} = \oint \vec{E} \cdot d\vec{A} = \frac{q^{enc}}{\varepsilon_0}$$

where q^{enc} is the net charge in coulombs enclosed by any chosen Gaussian surface and ε_0 is the the electric constant in a vacuum or air ($\varepsilon_0 = 1/4\pi k$ = $8.85 \times 10^{-12}\ \mathrm{C^2/N \cdot m^2}$). Gauss' law is typically used to compute the electric field at some distance from a uniform charge distribution that is symmetric. We will be interested in two types of symmetry—cylindrical symmetry and spherical symmetry. Learning to apply Gauss' law will take some practice. The key is to pick a closed surface with the same symmetry as the electric field associated with a charge distribution. Thus, a spherical surface works for point charges and spherical distributions and a cylindrical surface works for line charges and cylindrical distributions.

Geometry Review: Circles, Spheres, and Cylinders

Before we wade into the use of Gauss' law, let's review some geometry for some fairly simple symmetric shapes.

20.9.1. Activity: Some Geometry of Circles and Spheres

a. What is the equation for the circumference of a circle of radius r? If the radius doubles, what happens to the circumference?

b. What is the equation for the area of a circle of radius r? If the radius doubles, what happens to the area?

c. What is the equation for the volume of a sphere of radius r? If the radius doubles, what happens to the volume?

d. Find the derivative of the volume V as a function of r. Show how this derivative can be used to determine how much the volume of a sphere would increase (that is, the factor dV) if the radius of the sphere were increased from r to $r + dr$. **Hint:** Consider this increase as being the volume of a shell of thickness dr surrounding the sphere.

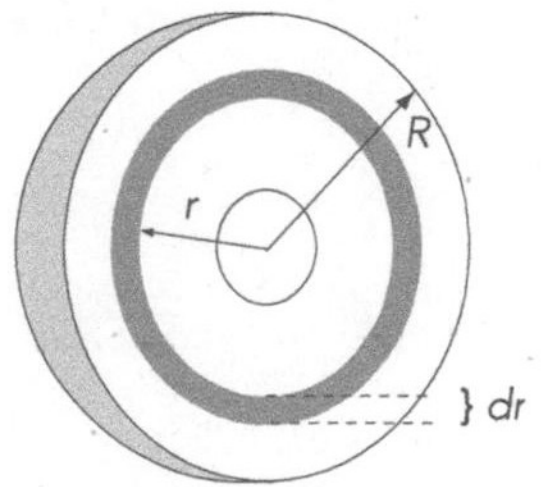

Fig. 20.8.

e. If the letter S is used to represent the surface area of a sphere, what is the volume dV of a thin shell of thickness dr that surrounds a sphere of radius r in terms of S and dr?

f. Use the derivative dV/dr from part d. and the idea that a spherical shell represents a volume increase to show that the surface area of a sphere can be represented by the equation $S = 4\pi r^2$.

g. If the radius of a sphere doubles, how much does its surface area increase?

h. If a charge Q is spread uniformly throughout the volume of an *insulating* sphere (that is, charge cannot move around inside it) of radius r, what fraction of the charge lies within a radius of $r/2$? **Warning:** The answer is not 1/2.

20.9.2. Activity: Some Geometry of Cylinders

a. Consider a cylinder of radius r and length L. What is its volume in terms of π, r, and L?

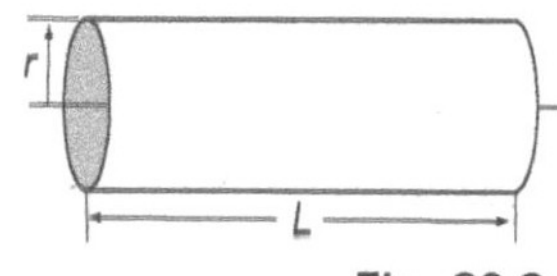

Fig. 20.9.

b. If a charge Q is spread uniformly throughout the volume of a cylinder of radius r and length L, what fraction of the charge lies within a radius of $r/2$? **Warning:** The answer is not 1/2.

c. What is the surface area of the cylinder in terms of π, r, and L? **Hint:** Don't neglect the ends.

Spherical Symmetry, Charge Density, and Gauss' Law

After the geometry review you just completed, you should be ready to use Gauss' law to find the equation describing the electric field at a distance r from a point charge. You should also be able to derive an equation for the electric field at a distance r from an insulator with a uniform, spherically symmetric charge distribution. You can find both these derivations in almost any calculus-based introductory physics text.

The concept of *charge density* is very useful in figuring out how much charge is contained within a given radius r in a charged sphere. Before you get started with your derivations of E-field equations, take a moment to read about the concept of charge density.

If a volume element dV contains a small amount of charge dq, then the charge density is given by the equation:

$$\rho = \frac{dq}{dV}$$

Thus, if the charge is distributed in a spherically symmetric manner, the amount of charge contained within a radius r is given by:

$$q = \int_0^r \rho\, dV = \int_0^r \rho\, 4\pi r^2\, dr = 4\pi \int_0^r \rho\, r^2 dr$$

In the case where the charge is uniformly distributed, ρ is a constant given by:

$$\rho = \frac{3Q}{4\pi R^3}$$

where Q is the total charge in the sphere and R is the sphere's radius. In this case, then, the charge contained within a radius r is given by:

$$q = \int_0^r \rho\, dV = 4\pi\rho \int_0^r r^2\, dr$$

Gauss' Law and the Point Charge

Let's begin by using Gauss' law in the form

$$\Phi^{net} = \oint \vec{E} \cdot d\vec{A} = \frac{q^{enc}}{\varepsilon_0}$$

to find the electric field magnitude at any distance, r, from a point charge $+q$.

20.9.4. Activity: Gauss' Law and a Point Charge

a. Write down the general expression for Gauss' law.

b. Explain why the dot product $\vec{E} \cdot d\vec{A}$ can be replaced with the expression EdA inside the integral where E is the magnitude of the electric field and dA is the magnitude of an element of area on the surface of the spherical shell. **Hint:** What is the angle between the vector $\vec{E}$ and the vector $d\vec{A}$ at any point on the spherical surface?

c. Explain why the magnitude of the electric field is the same at all points on the spherical shell. **Hint:** Think of Coulomb's law and the definition of electric field.

d. If the magnitude of the electric field, E, is the same at all points on the spherical surface, explain why E can be factored out of the integral $\int \vec{E} \cdot d\vec{A}$.

e. Since $4\pi r^2$ is the equation for the area of the surface of a sphere, what is the equation for $\int dA$?

f. Finally, what is the magnitude of the electric field, E, as a function of the central charge, q, and the distance from it, r?

Gauss' Law and a Spherically Symmetric Charge Distribution

Use some of the ideas from the last two activities about charge density and about using Gauss' law to find the magnitude of the electric field near a point charge to find the values of the magnitude of electric field in the vicinity of a spherically symmetric distribution of charges.

In the next activity you are to compute the magnitude of the electric field at a distance r from the center of a charged sphere of radius R with a total excess charge of q distributed uniformly throughout its volume, where $r < R$ (that is, the point in question is inside the sphere!).

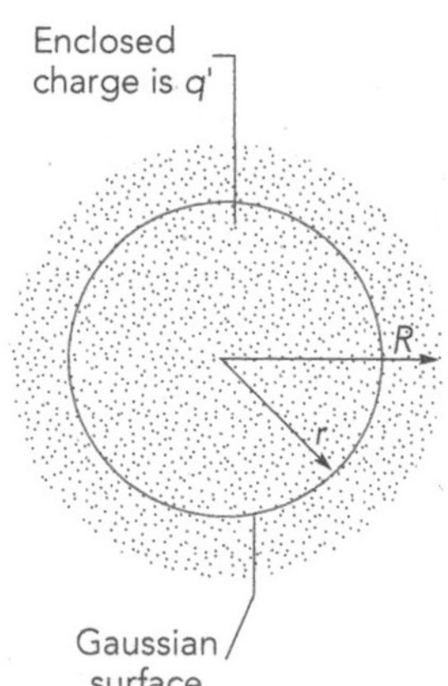

Fig. 20.10.
This figure is used by permission of John Wiley and Sons, Inc. and originally appeared in *Understanding Physics*, Cummings et. al., copyright © 2004 John Wiley and Sons.

20.9.4. Activity: Gauss' Law and Spherical Symmetry

a. To use Gauss' Law to find the electric field, you'll need to know the charge enclosed inside a sphere of radius r in which charges are distributed uniformly through the larger sphere of radius R. Start by using the equation for the volume of a sphere of radius r ($V = 4/3\,\pi r^3$) to show that the smaller amount of excess charge q' that lies within a radius r is given by

$$q' = q\,\frac{r^3}{R^3}$$

b. Next, use Gauss's law along with the equation you just derived to show that the magnitude of the electric field *inside* a uniformly charged sphere of radius R having a total or net excess charge of q is

$$E = |\vec{E}| = k\left(\frac{|q|r}{R^3}\right) \text{ where } k = \frac{1}{4\pi\varepsilon_0}$$

Gauss' Law and a Cylindrically Symmetric Charge Distribution

What happens to the electric field when there is cylindrical symmetry? Can Gauss' law be used to find the electric field at a distance r from an infinite line of electrical charges? How about the electric field at a distance r from an *infinitely* long insulator with a uniform charge distribution that has cylindrical symmetry? Consult an introductory text for hints if necessary.

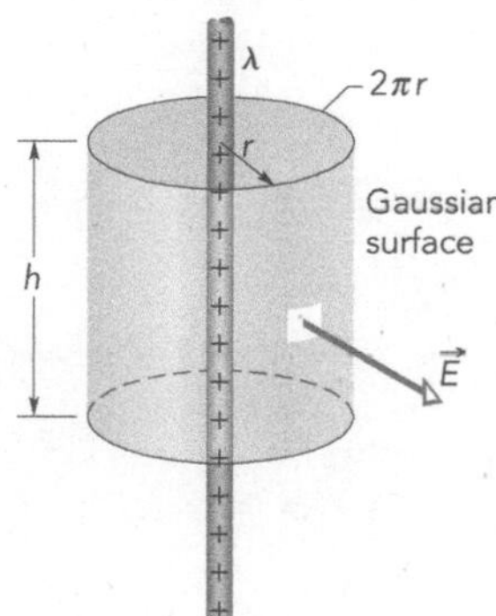

Fig. 20.11. A Gaussian surface in the form of a closed cylinder surrounds a section of a very long, uniformly charged, cylindrical plastic rod.

This figure is used by permission of John Wiley and Sons, Inc. and originally appeared in *Understanding Physics*, Cummings et. al., copyright © 2004 John Wiley and Sons.

20.9.5. Activity: Gauss' Law and Cylindrical Symmetry

a. Use Gauss' law to calculate the electric field at a distance r from a very long, straight, uniformly charged cylinder that has a charge per unit length of λ. (Mathematically, we can treat a wire which is *physically* very long as if it were infinitely long.) **Hint:** Using a symmetry argument, explain why you can neglect the electric field perpendicular to the two ends of the cylinder.

b. Assume that the charge per unit length $\lambda = 5.00 \times 10^{-9}$ C/m. Plug some numbers into the equation you derived to find the magnitude of the electric field at a distance of 5.00 cm from the line of charge. As usual, you should express your result using the appropriate number of significant figures.

Hint: $\varepsilon_0 = 8.85 \times 10^{-12} \mathrm{C^2/N \cdot m^2}$ (Electric constant)
$k = 1/(4\pi\varepsilon_0) = 8.99 \times 10^9 \mathrm{N \cdot m^2/C^2}$ (Coulomb constant)

c. What is the *direction* of the electric field? Explain the reasons for your answer.

d. How do we know that the *E*-field lines are perpendicular to the wire (that they have no component that is parallel to the wire)?

Name ______________________ Section ____________ Date ____________

UNIT 21: ELECTRICAL AND GRAVITATIONAL POTENTIAL

NGC 5457 is one of the most magnificent galaxies known to astronomers. What are the forces between elements of this galaxy? What forces does NGC 5457 exert on a small companion galaxy NGC 5474? It is amazing to think that if we know the masses of each element in these two galaxies, we can in principle calculate the gravitational potential energy of any mass element in the two galaxies. Although there are profound differences between electrical charge and mass, there is a similarity between the mathematical form of the electrical forces between charges and those between masses. Electrical potential energy is a key concept in understanding the behavior of the electric circuits that are such an important part of modern technology. In this unit you will study the mathematical concepts of electrical potential energy and voltage, and the similarity of these concepts to that of gravitational potential energy.

UNIT 21: ELECTRICAL AND GRAVITATIONAL POTENTIAL

I began to think of gravity extending to the orb of the moon, and . . . I deduced that the forces which keep the planets in their orbs must be reciprocally as the squares of their distances from the centers about which they revolve: and thereby compared the force requisite to keep the moon in her orb with the force of gravity at the surface of the earth, and found them to answer pretty nearly. All this was in the two plague years of 1665 and 1666, for in those days I was in the prime of my age for invention, and minded mathematics and philosophy more than at any time since. Isaac Newton

OBJECTIVES

1. To understand the similarities in the mathematics used to describe gravitational and electrical forces.

2. To review the mathematical definition of work and potential energy in a conservative force field.

3. To understand the definition of electrical potential, or *voltage,* and its similarity to the concept of gravitational potential.

4. To learn how to determine electric field lines from equipotential surfaces and vice versa.

5. To learn how to map equipotentials in a plane resulting from two po charges and two line electrodes.

21.1. OVERVIEW

The enterprise of physics is concerned ultimately with mathematically describing the fundamental forces of nature. Nature offers us several fundamental forces, which include a strong force that holds the nuclei of atoms together, a weak force that helps us describe certain kinds of radioactive decay in the nucleus, the force of gravity, and the electromagnetic force.

Two kinds of force dominate our everyday reality—the gravitational force acting between masses and the Coulomb force acting between electrical charges. The gravitational force allows us to describe mathematically how objects near the surface of the earth are attracted toward the earth and how the moon revolves around the earth and planets revolve around the sun. The genius of Newton was to realize that objects as diverse as falling apples and revolving planets are both moving under the action of the same gravitational force.

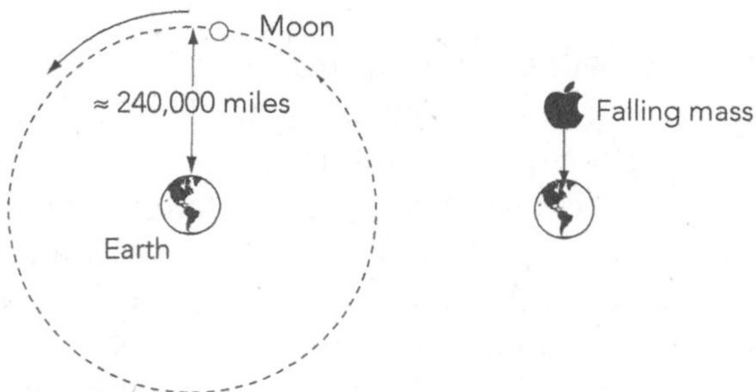

Fig. 21.1. Diagram showing that both the orbiting moon and a falling apple can experience a gravitational force that points toward the center of the Earth. The distance from the Earth to the moon is about 240,000 miles.

Fig. 21.2.

Similarly, the Coulomb force allows us to describe how one charge "falls" toward another or how an electron orbits a proton in a hydrogen atom.

The fact that both the Coulomb and the gravitational forces lead to object falling and to objects orbiting around each other suggests that these forc might have the same mathematical form.

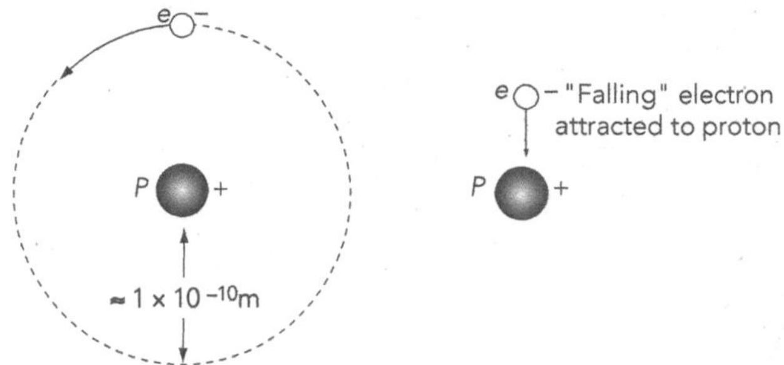

Fig. 21.3. Diagram showing that both an orbiting electron and a "fall ectron can experience a Coulomb force that points toward the center of the pr

In this unit we will explore the mathematical syr ween electrical and gravitational forces for two reasons. First, i to behold the unity that nature offers us since we use the same nematics to predict the motion of planets and galaxies, the fall s, the flow of electrons in circuits, and the nature of the hydr and of other chemical elements. Second, what you have already out the influence of the gravitational force on a mass and the conc ential energy in a gravitational field can be applied to aid your ur ing of the forces on charged particles. Similarly, a "gravitational" G can be used to find the forces on a small mass in the presence of a l erically symmetric mass.

We will introduce the concept of cal potential differences, which are analogous to gravitational potential y differences. An understanding of *electrical potential difference,* com called *voltage,* is essential to understanding the electrical circuits use nysics research and that dominate this age of electronic technology. Th will culminate in the actual measurement of electrical potential diffe s due to an electric field from two "line" conductors that lie in a plane a e mathematical prediction of the potential differences using a derivation d on Gauss' law.

ELECTRICAL AND GRAVITATIONAL FORCES

21.2. COMPARISON OF ELECTRICAL AND GRAVITATIONAL FORCES

Let's start our discussion of this comparison with the familiar expression of the Coulomb force exerted by charge B on charge A.

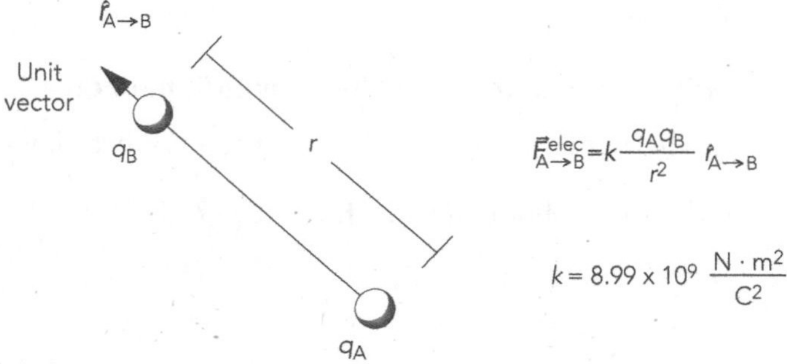

Fig. 21.4. This diagram shows how the direction of unit vector $\hat{r}_{A\to B}$ points in the direction from charge q_A to charge q_B.

Charles Coulomb did his experimental investigations of this force in the eighteenth century by exploring the forces between two small charged spheres. Much later, in the twentieth century, Coulomb's law enabled scientists to design cyclotrons and other types of accelerators for moving charged particles in circular orbits at high speeds.

Newton's discovery of the universal law of gravitation came the other way around. He thought about orbits first. This was back in the seventeenth century, long before Coulomb began his studies. A statement of Newton's universal law of gravitation describing the force experienced by mass 1 due to the presence of mass 2 is shown as follows in modern mathematical notation:

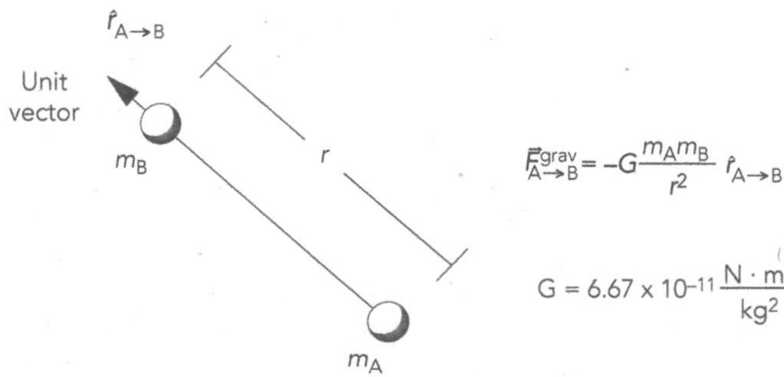

Fig. 21.5. This diagram shows how the direction of unit vector $\hat{r}_{A\to B}$ points from mass A to mass B.

About the time that Coulomb did his experiments with electrical charges in the eighteenth century, one of his contemporaries, Henry Cavendish, did a direct experiment to determine the nature of the gravitational force between two spherical masses in a laboratory. This confirmed Newton's gravitational force law and allowed him to determine the gravitational constant, G. A fact emerges that is quite amazing. *Both types of forces, electrical and gravitational, are very similar. Essentially the same mathematics can be used to describe orbital and linear motions due to either electrical or gravitational interactions of the tiniest fundamental particles or the largest galaxies.* This statement needs to be qualified a bit when electrons, protons, and other fundamental particles are considered. A new field called wave mechanics was developed in the early part of this century to take into account the wave nature of matter, which we don't actually study in introductory physics. However, even in wave mechanical calculations electrical forces like those shown above are used.

21.2.1. Activity: The Electrical vs. the Gravitational Force

Examine the mathematical expression for the two force laws.

a. What is the same about the two force laws?

b. What is different? For example, is the force between two like masses attractive or repulsive? How about two like charges? What part of each equation determines whether the like charges or masses are attractive or repulsive?

c. Do you think negative mass could exist? If there is negative mass, would two negative masses attract or repel?

21.3. WHICH FORCE IS STRONGER–ELECTRICAL OR GRAVITATIONAL?

Gravitational forces hold the planets in our solar system in orbit and account for the motions of matter in galaxies. Electrical forces serve to hold atoms and molecules together. If we consider two of the most common fundamental particles, the electron and the proton, how do their electrical and gravitational forces compare with each other?

Let's peek into the hydrogen atom and compare the gravitational force on the electron due to interaction of its mass with that of the proton to the electrical force between the two particles as a result of their charge. In order to do the calculation you'll need to use some well-known constants.

Electron: $m_e = 9.1 \times 10^{-31}$ kg $\quad q_e = -1.6 \times 10^{-19}$ C

Proton: $m_p = 1.7 \times 10^{-27}$ kg $\quad q_p = +1.6 \times 10^{-19}$ C

Distance between the electron and proton: $r \approx 1.0 \times 10^{-10}$ m

21.3.1. Activity: The Electrical vs. the Gravitational Force in the Hydrogen Atom

a. Calculate the magnitude of the electrical force on the electron. Is it attractive or repulsive?

b. Calculate the magnitude of the gravitational force on the electron. Is it attractive or repulsive?

c. Which is larger? By what factor (that is, what is the ratio)?

d. Which force are you more aware of on a daily basis? If your answer does not agree with that in part c, explain why.

21.4. GAUSS' LAW FOR ELECTRICAL AND GRAVITATIONAL FORCES

Gauss' law states that the net electric flux through any closed surface is equal to the net charge inside the surface divided by the electric constant in a vacuum, ε_0, where $\varepsilon_0 = 1/(4\pi k)$. Mathematically this is represented by an integral of the dot product of the electric field and the normal to each element of area over the closed surface:

$$\Phi^{net} = \oint \vec{E} \cdot d\vec{A} = \frac{q^{enc}}{\varepsilon_0}$$

The fact that electric field lines spread out so that their density (and hence the strength of the electric field) decreases at the same rate that the area of an enclosing surface increases can ultimately be derived from the $1/r^2$ dependence of electrical force on distance. Thus, *Gauss' law should also apply to gravitational forces.*

21.4 vity: The Gravitational Gauss' Law

gravitational version of Gauss' law in words and then represent
ı an equation. **Hint:** The electric field vector, $\vec{E}$, is defined as where is the force that source charges exert on a tiny test charge ed at a point in space of interest. Let's define the gravitational l vector $\vec{Y}$ as $\vec{F}^{grav}/m$ by analogy.

V use the new version of Gauss' law to calculate the gravitational field distance from the surface of the earth just as we can use the electrical law to determine the electric field at some distance from a uniformly d sphere. This is useful in figuring out the familiar force "due to grav- ear the surface of the earth and at other locations.

1.4.2. Activity: The Gravitational Force of the Earth

a. Use Gauss' law to show that the magnitude of the gravitational field, $\vec{Y}$, at a height h above the surface of the earth is given by $GM/(R+h)^2$ where M is the Earth's mass and R is the radius. **Hints:** How much mass is enclosed by a spherical shell of radius $R+h$? Does $\vec{Y}$ have a constant magnitude everywhere on the surface of the spherical shell? Why? If so, can you pull it out of the Gauss' law integral?

b. Calculate the gravitational field, $\vec{Y}$, at the surface of the earth where $h = 0.00$ m. Assume the radius of the earth is $R \approx 6.38 \times 10^3$ km and its mass $M \approx 5.98 \times 10^{24}$ kg. Does the result look familiar? How is $\vec{Y}$ related to the local gravitational constant given by $g = 9.8$ m/s^2?

c. Use the equation you derived in part a to calculate the value of $\vec{Y}$ at the ceiling of the room you are now in. How does it differ from the value of $\vec{Y}$ at the floor? Can you measure the difference in the lab using the devices available?

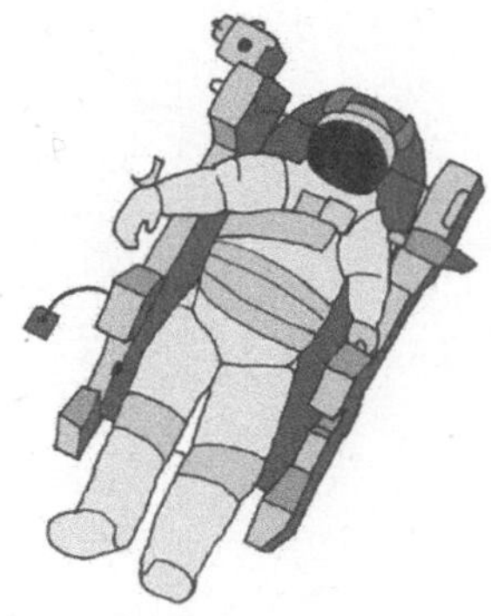

Fig. 21.6.

d. Suppose you travel halfway to the moon. What is the new value of $\vec{Y}$? Can you measure the difference? (Note that the earth-moon distance is about 384,000 km.)

e. Is the local gravitational "constant," g, really a constant? Explain.

f. In part d, you showed that there is a significant gravitational attraction halfway between the earth and the moon. Why, then, do astronauts experience weightlessness when they are orbiting a mere 120 km above the earth?

Fig. 21.7.

21.5. WORK IN A GRAVITATIONAL FIELD–A ' IEW

Let's review some old definitions in preparation ackling the idea of work and energy expended in moving through an ele field. Work, like flux, is a scalar rather than a vector quantity. It is define thematically as a line integral over a path between locations 1 and 2. E element of $F \cdot d\vec{s}$ represents the dot product or projection of $\vec{F}$ along $d\vec{s}$ at n place along a path.

$$W = \int_1^2 \vec{F} \cdot d\vec{s} \text{ where } \vec{F} \cdot \quad F\,(ds)\cos\theta$$

In this general situation, the angle θ betw $\vec{F}$ and $d\vec{s}$ may be different at every location along the chosen path.

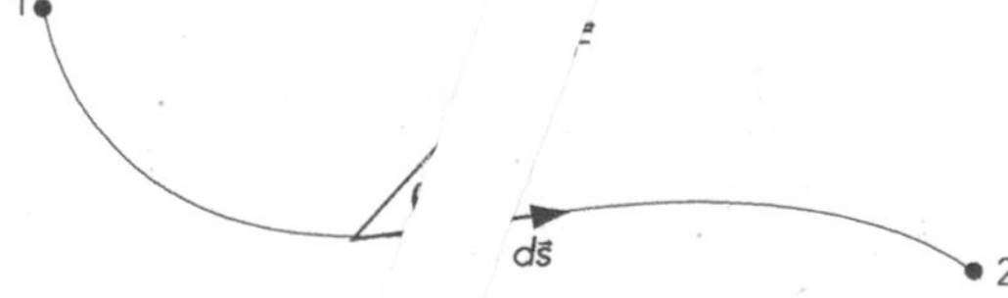

Fig. 21.8. This diagram shows a curve th from point 1 to point 2. At one point along the path, a small test mass cou perience a force $\vec{F}$ that is in a different direction than a small path element $d\vec{s}$.

Let's review the procedur r calculating work and for determining that the work done in a conserva force field is independent of path by taking some work measurements a two different paths, *ac* and *adc*, as shown in Fig. 21.9.

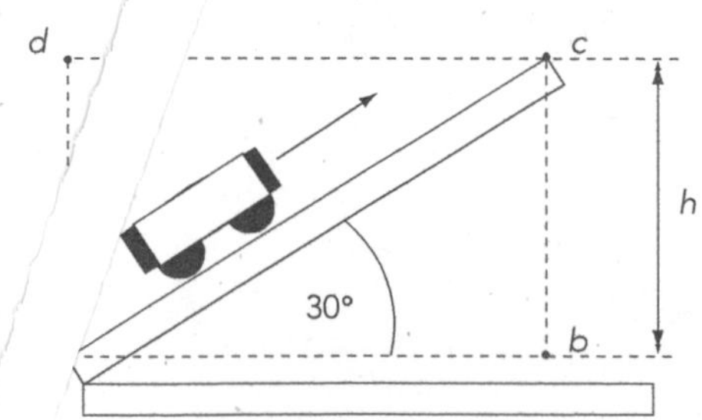

Fig. 21.9. Diagram of a mass moving up a 30° incline.

In order to take measurements you will need the following equipment:

- 1 inclined plane
- 1 low-friction cart
- 1 metric spring scale, 5 N
- 1 meter stick
- 1 protractor

Recommended group size:	2	Interactive demo OK?:	N

Set the inclined plane to an angle of about 30° to the horizontal and take any necessary length measurements. Don't forget to list your units!

21.5.1. Activity: Earthly Work

a. Use path *ac* to raise the cart up an inclined plane that makes an angle of about 30 degrees with the earth's surface (see diagram above). Use a spring balance to measure the force you are exerting, and calculate the work you performed *using the definition of work.*

b. Raise the cart directly to the same height as the top of the inclined plane along path *ad* and then move the cart horizontally from point *d* to point *c*. Again, use the *definition of work* to calculate the work done in raising the cart along path *adc.*

c. How does the work measured in a compare to that measured in b? Is this what you expected? Why or why not? **Hint:** Is the force field conservative?

21.6. WORK AND THE ELECTRIC FIELD

It takes work to lift an object in the earth's gravitational field. Lowering the object releases the energy that was stored as potential energy when it was lifted. Last semester, we applied the term *conservative* to the gravitational force because it "releases" *all* of the stored energy. We found experimentally that the work required to move a mass in the gravitational field was path independent. This is an important property of any conservative force. Given the mathematical similarity between the Coulomb force and the gravitational force, it should come as no surprise that experiments confirm that an electric field is also conservative. This means that the work needed to move a charge from point A to point B is independent of the path taken between points. A charge could be moved directly between two points or looped around and the work expended to take either path would be the same. Work done by an electric field on a small test charge q_t traveling between points A and B is given by

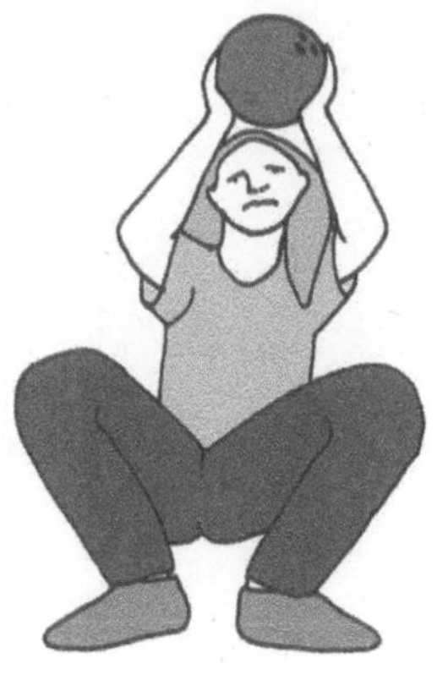

Fig. 21.10.

$$W = \int_A^B \vec{F}^{elec} \cdot d\vec{s} = \int_A^B q_t \, \vec{E} \cdot d\vec{s}$$

21.6.1. Activity: Work Done on a Charge Traveling in a Uniform Electric Field

a. A test charge q_t travels a distance d from point A to point B; the path is parallel to a uniform electric field of magnitude E. What is the work done by the field on the charge? How does the form of this equation compare to the work done on a mass m traveling a distance d in the almost uniform gravitational field near the surface of the earth?

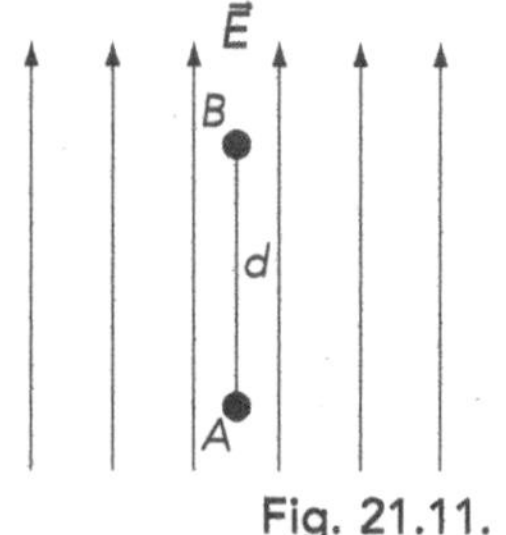

Fig. 21.11.

b. The small test charge q_t travels a distance d from point A to point B in a uniform electric field of magnitude E, but this time the path is perpendicular to the field lines. What is the work done by the field on the charge?

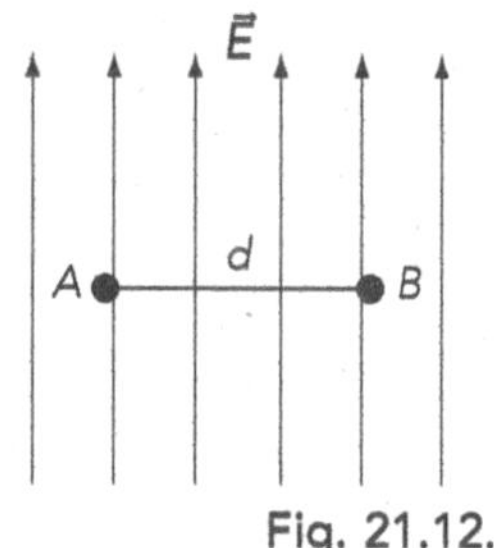

Fig. 21.12.

c. The test charge q_t travels a distance d from point A to point B in a uniform electric field of magnitude E. The path lies at a 45° angle to the field lines. What is the work done by the field on the charge?

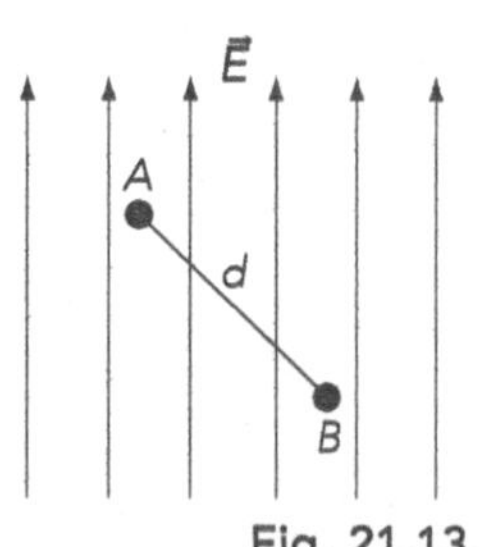

Fig. 21.13.

ELECTRIC POTENTIAL DIFFERENCE

21.7. POTENTIAL ENERGY AND POTENTIAL DIFFERENCE

Recall that by definition the work done by a conservative force equals the negative of the change in potential energy ($W^{cons} = -\Delta U^{cons}$). If electrical forces are conserved, the change in potential energy of a charge moving from point A to point B under the influence of an electrical force is given by $\Delta U^{elec} = -W^{elec}$:

so that
$$\Delta U^{elec} = U_B^{elec} - U_A^{elec} = -\int_A^B q_t \vec{E} \cdot d\vec{s}$$

By analogy to the definition of the electric field, we are interested in defining the *electric potential difference* $\Delta V = V_B - V_A$ as the change in electrical potential energy ΔU^{elec} per unit charge. Formally, *the potential difference is defined as the work per unit charge that an external agent must perform to move a test charge from A to B without changing its kinetic energy.* The potential difference has units of joules per coulomb. Since 1 J/C is defined as *one volt,* the potential difference is often referred to as *voltage.*

21.7.1. Activity: The Equation for Potential Difference

Write the equation for the potential difference as a function of $\vec{E}$, $d\vec{s}$, A, and B.

21.8. THE POTENTIAL DIFFERENCE FOR A POINT CHARGE

The simplest charge configuration that can be used to consider how voltage changes between two points in space is a single point charge. We will start by considering a single point charge and then move on to more complicated configurations of charge.

A point charge q produces an electric field that moves out radially in all directions. The line integral equation for the potential difference can be evaluated to find the potential difference between any two points in space A and B.

It is common to choose the reference point for the determination of voltage to be at infinity, denoted ∞, so that we are determining the work per unit charge that is required to bring a test charge from infinity to a certain point in space. Let's choose a coordinate system so that the point charge is conveniently located at the origin. In this case we will be interested in the potential difference between infinity and some point that is a distance r from the point charge. Thus, we can write the equation for the potential difference, or voltage, as

$$\Delta V = V_B - V_A = V_r - V_\infty = -\int_\infty^r \vec{E} \cdot d\vec{s}$$

Often, when the reference point for the potential difference is at infinity, this difference is simply referred to as "the potential" and the symbol ΔV is just replaced with the symbol V.

21.8.1. Activity: Potential at a Distance r from a Charge

Show that, if A is at infinity and B is a distance r from a point-like charge q, then the potential V is given by the expression $V = kq/r$. **Hint:** What is the mathematical expression for an E-field from a point charge?

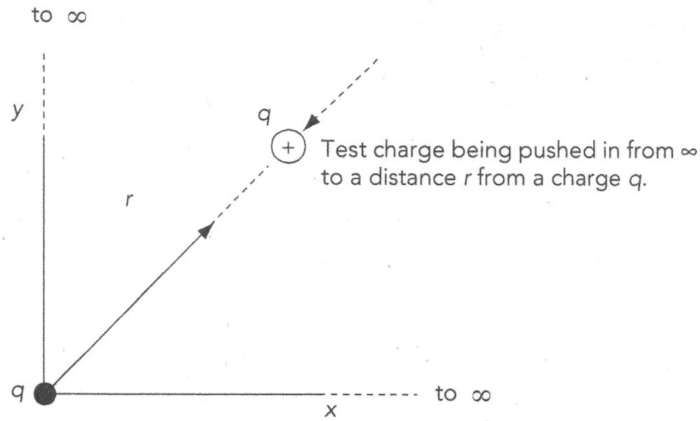

Fig. 21.14. Charge at the origin.

21.9. THE POTENTIAL DIFFERENCE DUE TO CONTINUOUS CHARGE DISTRIBUTIONS

The potential from a continuous charge distribution can be calculated several ways. Each method should yield approximately the same result. First, we can use an integral method in which the potential dV from each element of charge dq is integrated mathematically to give a total potential at the location of interest. Second, we can approximate the value of the potential V by summing up several finite elements of charge Δq by using a computer spreadsheet or hand calculations. Finally, we can use Gauss' law to find the electric field along with the defining equation for potential difference to set up the appropriate line integral shown in Activity 21.7.1.

Again, let's consider a relatively simple charge distribution. In this case we will look at a ring with charge uniformly distributed on it. We will calculate the potential on the axis passing through the center of the ring as shown in the diagram below. (Later on you could find the potential difference from a disk or a sheet of charge by considering a collection of nested rings.)

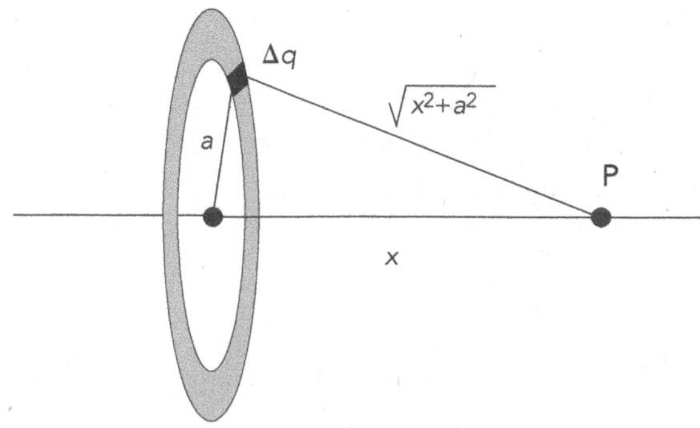

Fig. 21.15. Diagram showing the distance between an element on a charged ring of radius r and point P along an axis passing through the center of the ring.

A ring of charge has a total charge of $q^{tot} = 20\ \mu C$ (or 20×10^{-6} C). The radius of the ring, a, is 30 cm. What is the electric field, $\vec{E}$, at a distance of x cm from the ring along an axis that is perpendicular to the ring and passing through its center? What is the potential, V? Let's begin by calculating the potential. **Hints:** Since the potential is a scalar and not a vector, we can calculate the potential at point P (relative to ∞) for each of the charge elements Δq and add them to each other. This looks like a big deal but it is actually a trivial problem because all the charge elements are the same distance from point P.

21.9.1. Activity: Estimate of the Potential from a Charged Ring

a. Divide the ring into 20 elements of charge Δq and calculate the total V at a distance of $x = 20$ cm from the center of the ring using a spreadsheet program. Summarize the result below.

21.9.2. Activity: Calculation of the Potential from a Charged Ring

By following the steps below, you can also use an integral to find the exact value of the potential.

a. Show that $V = k\int \frac{dq}{r} = k\int \frac{dq}{\sqrt{x^2 + a^2}}$

b. Show that $k\int \frac{dq}{\sqrt{x^2+a^2}} = \frac{k}{\sqrt{x^2+a^2}}\int dq$ (that is, show that $\sqrt{x^2+a^2}$ is a constant and can thus be pulled out of the integral).

c. Perform the integration in part b. above. Then substitute values for the ring radius, a, the distance to point P, x, and the total charge on the ring, q^{tot}, into the resulting exprcssion in order to obtain a more "exact" value for the potential.

d. How does the "numerical" value that you obtained in Activity 21.9.1 compare with the "exact" value you obtained in c?

Now let's take a completely different approach to this problem. If we can find the vector equation for the electric field at point P due to the ring of charge, then we can use the expression

$$\Delta V = V(x) - V_{\infty} = -\int_{\infty}^{x} \vec{E} \cdot d\vec{s}$$

where $V(x)$ is the electric potential relative to infinity on the x-axis a distance x from the center of the ring. This equation provides us with an alternative way to find a general equation for the potential at point P.

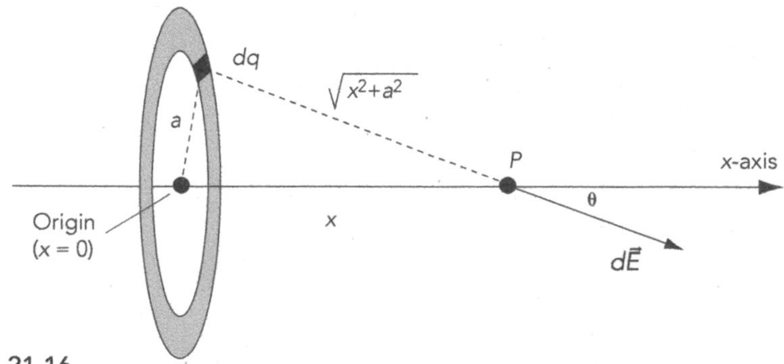

Fig. 21.16

21.9.3. Activity: ΔV from a Ring using the E-field Method

a. Show that the electric field at point P from the charged ring is given by

$$\vec{E} = \frac{kq^{tot}x}{(x^2+a^2)^{3/2}}\hat{\imath} = \frac{kq^{tot}x}{(\sqrt{x^2+a^2})^3}\hat{\imath}$$

Hints: 1. There is no y-component of the E-field on the x-axis. Why?

2. $\cos\theta = \dfrac{x}{\sqrt{x^2 + a^2}}$

b. Use the equation $\Delta V = V(x) - V_\infty = -\int_\infty^x \vec{E} \cdot d\vec{s}$ to show that $\Delta V = \dfrac{kq^{\text{tot}}}{\sqrt{x^2 + a^2}}$.

c. Use the equation you verified in part b. to find a numerical value for ΔV.

d. How does the result compare to that obtained in Activity 21.9.2c?

21.10. EQUIPOTENTIAL SURFACES

Sometimes it is possible to move along a surface without doing any work. Thus, it is possible to remain at the same potential energy anywhere along such a surface. If an electric charge can travel along a surface without doing any work, the surface is called an *equipotential surface.*

Consider the three different charge configurations shown below. Where are the equipotential surfaces? What shapes do they have? **Hint:** If you have any computer simulations available to you for drawing equipotential lines associated with electrical charges, you may want to check your guesses against the patterns drawn in one or more of the simulations.

21.10.1. Activity: Sketches of Electric Field Lines and Equipotentials

a. Suppose that you are a test charge and you start moving at some distance from the charge below (such as 4 cm). What path could you move along without doing any work—that is, $\vec{E} \cdot d\vec{s}$ is always zero? What is the shape of the equipotential surface? Remember that in general you can move in *three* dimensions.

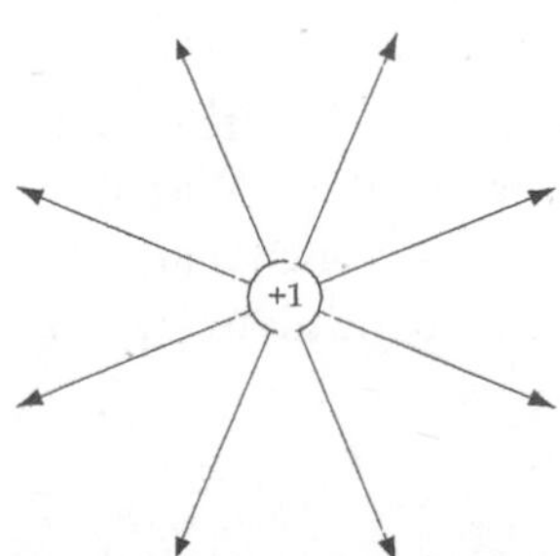

b. Find some equipotential surfaces for the charge configuration shown below, which consists of two charged metal plates placed parallel to each other. What is the shape of the equipotential surfaces?

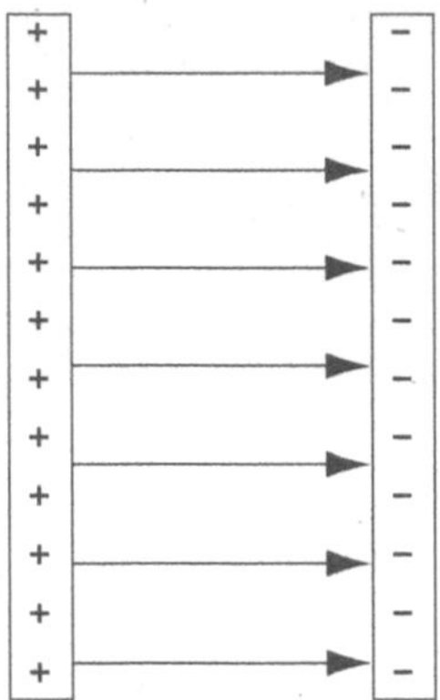

c. Find some equipotential surfaces for the electric dipole charge configuration shown below.

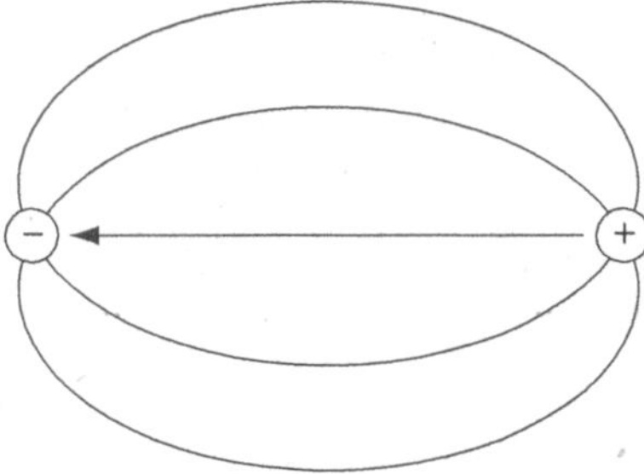

d. In general, what is the relationship between the direction of the equipotential lines you have drawn (representing that part of the equipotential surface that lies in the plane of the paper) and the direction of the electric field lines?

21.11. EXPERIMENT ON EQUIPOTENTIAL PLOTTING

The purpose of this observation is to explore the pattern of potential differences in the space between conducting equipotential surfaces. This pattern of potential differences can be related to the electric field caused by the charges that lie on the conducting surfaces.

You will be using pieces of carbonized paper with conductors painted on them in different shapes to *simulate* the pattern of equipotential lines associated with metal electrodes in air. (**Warning:** This is only a simulation of "reality"!) One of the papers has two small circular conductors painted on it and the other has two linear conductors.

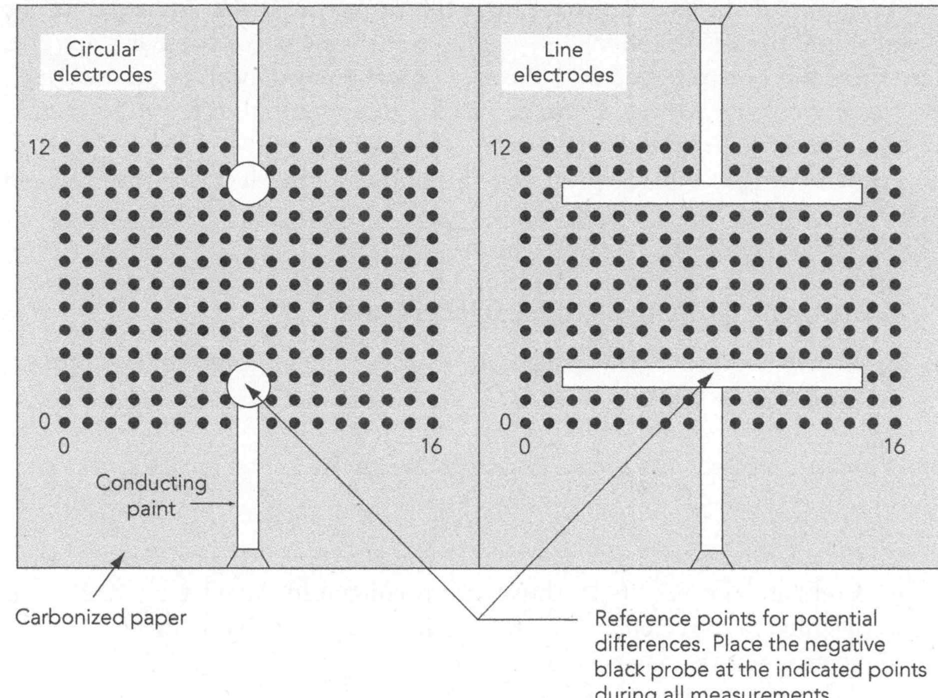

Fig. 21.17. Equipotential plotting sheets consisting of carbonized papers with point and line electrodes painted on them with conducting paint.

You can use a battery to set up a voltage (i.e., a potential difference) across a paper with circular electrodes. You can then use a digital voltmeter to trace several equipotential lines on it. To do this observation you will need the following equipment:

- 2 equipotential plotting sheets (with line and circular electrodes)
- 1 digital voltmeter
- 1 battery, 5 V
- 2 alligator clip leads

Recommended group size:	4	Interactive demo OK?:	N

Pick a piece of paper with either the two circular electrodes or the line electrodes. Use the alligator clips to connect the terminals of the battery to each of the electrodes on the paper as shown below.

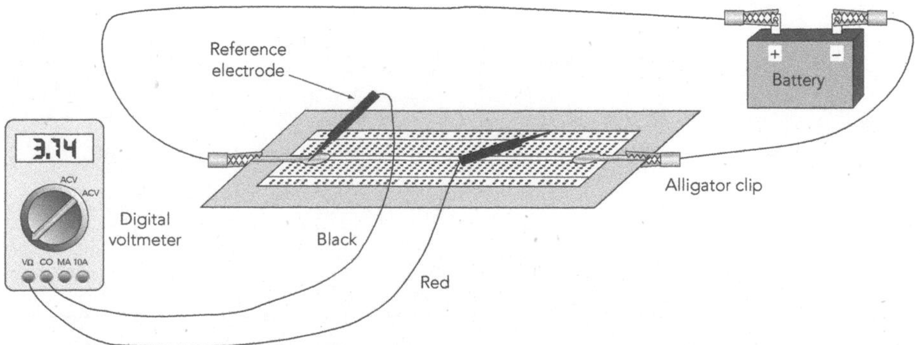

Fig. 21.18. Suggested set-up for measuring potential differences between various points on equipotential plotting sheets.

Turn on the digital voltmeter. Set the tip of the black voltmeter probe (plugged into the COM input) on the center of the negative circular electrode. Place the red probe (plugged into the V-Ω input) on any location on the paper. The reading on the voltmeter will determine the potential difference between the two points. What happens when you reverse the probes? Why?

After placing the black probe on the center of an electrode, use the red probe to find the equipotential lines for potential differences of $+1v$, $+2v$, $+3v$, and $+4v$.

21.11.1. Activity: Mapping the Equipotentials

a. Sketch the lines you mapped to scale on the appropriate diagram on the previous page.

b. Compare the results to those you predicted in Activity 21.10.1. Is it what you expected? If not, explain how and why it differs. Make sketches in the space below, if that is helpful.

Note: Be sure to turn the voltmeter *off* when you are finished with your observations.

Name ______________________ Section __________ Date __________

UNIT 22: BATTERIES, BULBS, AND CURRENT FLOW

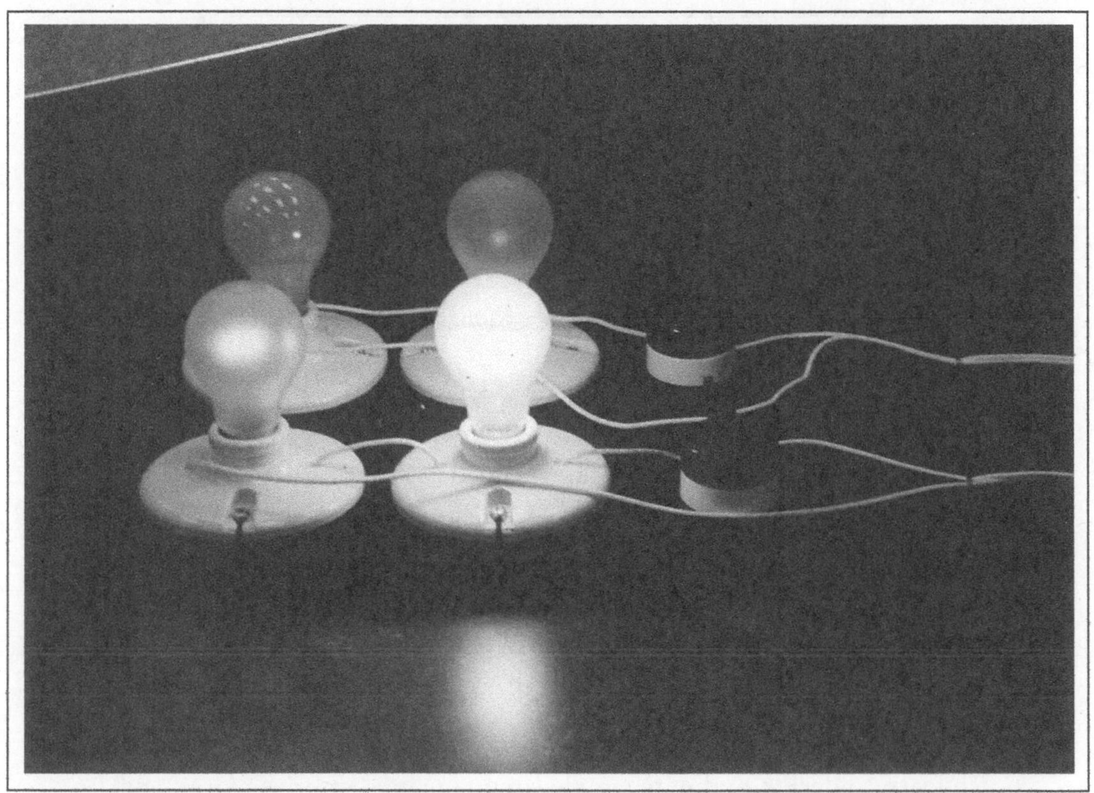

The two front bulbs on this light board are wired together in a standard "parallel" arrangement. When they are plugged into a 110 v.a.c. outlet in a house, the 150-watt bulb on the right is much brighter than the 75-watt bulb on the left. What happens to the bulbs if a very subtle wiring change is made so that they are connected in series? Which one is brighter now? Do they combine to give a reader as much illumination as before? Most people find what happens startling. When you finish working with different combinations of batteries and bulbs in this unit you ought to be able to predict the outcome of this change from parallel to series wiring.

UNIT 22: BATTERIES, BULBS, AND CURRENT FLOW*

We have in us somewhere knowledge . . . of the taste of strawberries. . . . When we bite into a berry, we are ready *to taste a certain kind of taste; if we taste something very different, we are surprised. It is this—what we expect or what surprises us—that tells us best what we really know.*

John Holt

OBJECTIVES

1. To understand how a potential difference results in a current flow through a conductor.
2. To learn to design and wire simple circuits using batteries, wires, and switches.
3. To learn to use symbols to draw circuit diagrams.
4. To learn to use an ammeter for measuring current and a voltmeter for measuring voltage.
5. To understand the relationship between the current flows in series and parallel circuits.
6. To understand the concept of resistance.

* Portions of this unit are based on research by Lillian C. McDermott and Peter S. Shaffer published in *AJP 60,* 994–1012 (1992).

22.1. OVERVIEW

Fig. 22.1.

In the following sessions, we are going to discover, extend, and apply theories about electric charge and potential difference to electric circuits. The study of circuits will prove to be one of the more practical parts of the whole introductory physics course sequence, since electric circuits form the backbone of much of twentieth-century technology. Without circuits we wouldn't have electric lights, air conditioners, automobiles, telephones, TV sets, dishwashers, computers, Xerox® machines, or electric toothbrushes.

In the last unit you used a battery to establish a potential difference (or voltage) across two electrodes. *Battery* is a term applied to any device that generates an electrical potential difference from other forms of energy. The type of batteries you are using in this course are known as chemical batteries because they convert internal chemical energy into electrical energy.

As a result of a potential difference, an electrical charge can be repelled from one terminal of the battery and attracted to the other. No charge can flow out of a battery unless there is a conducting material connected between its terminals. A flow of charge can cause a small light bulb to glow.

In this unit, you are going to explore how charge originating in a battery flows in wires and bulbs. You will be asked to develop and explain some models that predict how the charge will flow in series and parallel circuits. You will also be asked to devise ways to test your models using an ammeter to measure the rate of flow of electrical charge through it and using a digital voltmeter to measure the potential difference between two points in a circuit.

CURRENT FLOW

22.2. WHAT IS ELECTRIC CURRENT?

We have attributed the forces between objects that are rubbed in particular ways to a property of matter known as charge. Most textbooks assert that the electric current that flows through the wires connected to a battery are charges in motion. How do we know this? Perhaps "current" is something else—another phenomenon. This is a question that received a great deal of attention from Michael Faraday, a famous early nineteenth-century scientist. As a result of his researches, including electric eel experiments, Faraday produced a table like the one shown below. He concluded that "electricity, whatever may be its source, is identical in its nature."*

Reproduction of Faraday's Table. The X's denote results obtained by Faraday and the +'s denote positive results found by other investigators later.

	Physiological effect	Magnetic deflection	Magnets made	Spark	Heating power	True chemical action	Attraction and repulsion	Discharge by hot air
1. Voltaic electricity	X	X	X	X	X	X	X	X
2. Common electricity	X	X	X	X	X	X	X	X
3. Magneto-electricity	X	X	X	X	X	X	X	
4. Thermo-electricity	X	X	+	+	+	+		
5. Animal electricity	X	X	X	+	+	X		

Comparing Stuff from a Battery to the Rubbing Stuff

By using the following apparatus you can do some of your own comparisons. You'll need the following:

- 2 aluminum angle irons, approx. 15 cm
- 1 metal-coated, threaded ball (with low mass)
- 1 battery pack or power supply with approximately 300 volts
- 1 black plastic rod
- 1 fur
- 1 glass rod

* Faraday, M. "Identity of Electricities Derived from Different Sources," in *Experimental Researches in Electricity, Vol. I,* Taylor and Francis, London. (Reprinted by Dover Publications, New York, 1965, p. 76.)

- 1 polyester cloth
- 1 electroscope
- 6 alligator clip leads
- 1 Wimshurst generator (optional)

Recommended group size:	4	Interactive demo OK?:	Y

Set up the electroscope and the angle irons (which will serve as what you will come to know as a capacitor) as shown in the following diagram.

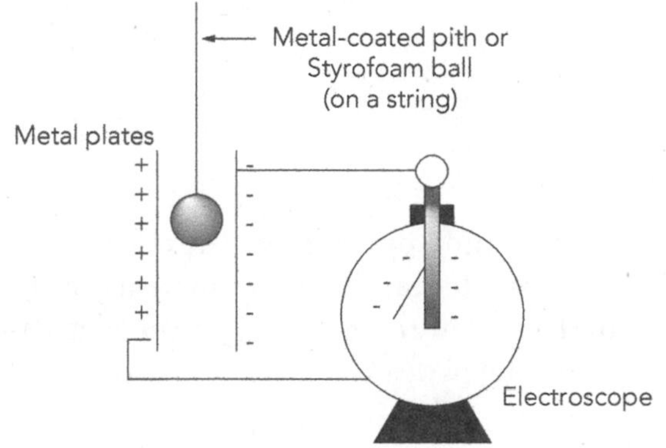

Fig. 22.2. Apparatus for detecting charge.

By "charging" the metal plates in two or more ways, you can test whether or not the different charging methods have different effects on the electroscope and on the ball dangling between metal plates. The main charging methods to be tested are:

Fig. 21.3.

1. *Electrostatic Charging by Rubbing*: Stroke one plate with a rubber rod that has been rubbed with the cat fur. Repeat this several times. Stroke the other plate with the glass rod that has been rubbed with polyester cloth.
2. *Charging with a Battery*: Connect a wire from the negative terminal of the battery pack to one of the plates. At the same time connect a wire from the positive terminal of the battery pack to the other plate.
3. *Charging with a Wimshurst Generator*: Connect a wire from one of the two terminals of the generator to one plate and a wire from the other terminal to the other plate.

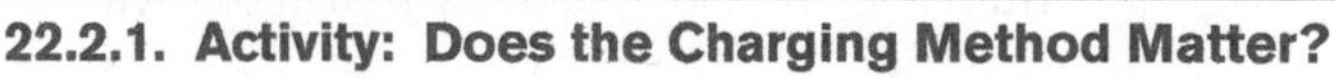

22.2.1. Activity: Does the Charging Method Matter?

a. Use the familiar rubbing method (method #1) to charge the plates. What happens to the electroscope? Why?

b. Again, use the rubbing method to charge the plates. Separate the metal plates so the gap between them is just barely bigger than the diameter of the foil-covered ball. Next, place the ball carefully between the metal plates. What happens? You should see something pretty unusual. In terms of the attraction and repulsion of different types of charges, explain why you see this unusual phenomenon.

c. Use the battery (and, optionally, the Wimshurst generator) to "charge" the plates. Repeat the tests from parts a. and b. What differences do you observe in the response of the electroscope and the ball to the charges on the plates?

d. Do the charges generated by rubbing and from the output of the battery cause different effects? If so, describe them. Do the charges generated in these two ways appear to be different?

The Official Mathematical Definition of Current

The rate of flow of electric charge is more commonly called electric current. If charge is flowing through a conductor, then the official mathematical definition of the average current is given by

$$<i> \equiv \frac{\Delta Q}{\Delta t} \qquad \text{(definition of average current)}$$

Instantaneous current is defined in the usual way with a limit:

$$i \equiv \lim_{\Delta t \to 0} \frac{\Delta Q}{\Delta t} = \frac{dQ}{dt} \qquad \text{(definition of instantaneous current)}$$

The unit of current is called the ampere (A). One ampere represents the flow of one coulomb of charge through a conductor in a time interval of one second.

22.3. LIGHTING A BULB

You can begin to explore circuits and currents by lighting a bulb with a battery. You will need:

- 1 #14 V bulb
- 1 D-cell battery, 1.5 V, alkaline
- 1 wire with alligator clip leads, > 10 cm
- common objects: paper clips, pencils, etc.

Recommended group size:	2	Interactive demo OK?:	N

Use the materials listed above to find some arrangements in which the bulb lights and some in which it does not light. For instance, does the bulb light up in the following arrangement?

Fig. 22.4. A wiring configuration that might cause a bulb to light in the presence of a battery.

22.3.1. Activity: Arrangements that Cause Light

a. Sketch two different arrangements in which the bulb lights.

b. Sketch two arrangements in which the bulb doesn't light.

c. Describe as fully as possible what conditions are needed if the bulb is to light and how these conditions are not satisfied in the arrangements that fail to cause the bulb to light.

Putting Other Objects Between the Battery and Bulb

Set up the single wire, battery, and bulb so that the bulb lights. Then, somewhere between the battery and the bulb, stick in a variety of material objects available in the room, such as paper, coins, rubber bands, fingers, pencils, keys, etc. Reflect on which types of materials allow the battery to light the bulb and which types don't. Since it seems that something flows from the battery to the bulb, we refer to materials that allow this flow as *conductors* and those that don't as *non-conductors.*

22.3.2. Activity: Materials That Allow the Bulb to Light

a. List some materials that allow the bulb to light.

b. List some materials that prevent the bulb from lighting.

c. What categories of materials are conductors? What categories seem to be non-conductors?

22.4. USING A BATTERY HOLDER, BULB SOCKET, AND SWITCH

Are you having trouble holding things together? Let's make it easier by using a battery holder and a bulb socket. While we're at it, let's also add a switch in the circuit. You will need:

- 1 #14 bulb
- 1 #14 socket
- 1 D-cell battery, 1.5 V, alkaline
- 1 D-cell holder
- 2 alligator clip leads, > 10 cm
- 1 SPST switch

Recommended group size:	2	Interactive demo OK?:	N

If you were to wire up the configuration shown in Figure 22.5, when would the bulb light? With the switch open (that is, so no contact between the wires is made)? Closed? Neither time?

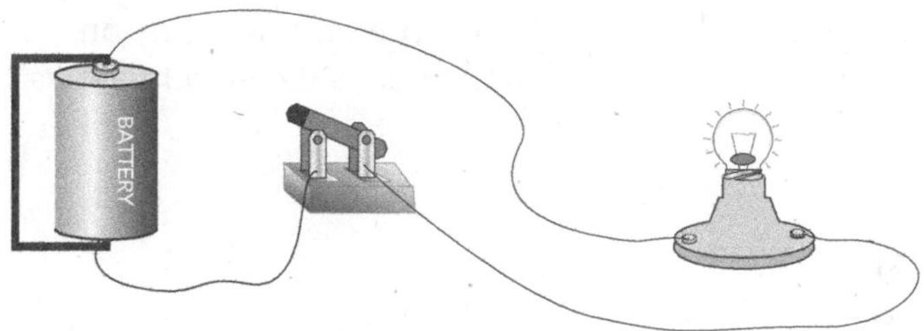

Fig. 22.5. A circuit with a battery, switch, and bulb holder.

22.4.1. Activity: Action of the Switch

a. Examine the bulb socket carefully. Explain what it does. What happens when you unscrew the bulb? Why doesn't the bulb light?

b. Examine the bulb closely. Use a magnifying glass, if available. Figure 22.6 shows the parts of the bulb that are hidden from view. Why is the filament of the bulb connected in this way?

Fig. 22.6. Wiring inside a light bulb.

c. Will the bulb light with the switch open (no contact), closed (contact), or neither time? Explain the reasons for your prediction from your previous work.

d. Wire the circuit in Figure 22.5, test it, and describe what happens.

e. Leave the switch closed so that the bulb remains on for 5 to 10 seconds. Feel the bulb. What happens as current flows through the bulb?

f. What characteristics must the current path have for the bulb to glow?

22.5. DESIGNING YOUR OWN ELECTRICAL DEVICES

Let's apply what you know about circuits to make some original devices. You can use extra switches, wires, bulbs, etc., if needed. Construct all three of the devices described below. Sketch the circuit for each of your devices. You can use the following equipment:

- 3 #14 bulbs
- 3 #14 bulb holders
- 1 D-cell battery, 1.5 V
- 1 D-cell holder
- 6 alligator clip leads, > 10 cm
- 1 SPST switch
- 1 DPDT switch
- 1 SPDT switch

Recommended group size:	2	Interactive demo OK?:	N

1. *Christmas Tree Lights*: Suppose you want to light up your Christmas tree with three bulbs. What happens if a bulb fails? (Don't break the bulb! You can simulate failure by loosening a bulb in its socket.) Figure out a way to wire in all three bulbs so that the other two will still be lit if any one of the bulbs burns out.

Fig. 22.7.

2. *Lighting a Tunnel*: The bulbs and switches must be arranged so that a person walking through a tunnel can turn on a lamp for a part of the tunnel and then turn on a second lamp in such a way that the first one turns off automatically.
3. *Caller Indicator for the Deaf*: A deaf person should be able to see, by looking at one or two bulbs, whether a visitor is at the front or back door of a house. You should do this with only 1 battery, 2 bulbs, and 2 switches.

22.5.1. Activity: Description of the Invention(s)

a. Include a description and a circuit drawing for your Christmas tree lights.

b. Include a description and a circuit drawing for your tunnel lighter.

c. Include a description and a circuit drawing for your deaf caller. It should have no more than one battery.

22.6. CIRCUIT DIAGRAMS

Now that you have been wiring circuits and drawing them, you may be getting tired of drawing pictures of the batteries, bulbs, and switches in your circuits. A series of symbols have been created to represent circuits. These symbols will enable you to draw the nice neat square-looking circuit diagrams that you see in physics textbooks. A few of the electric circuit symbols follow.

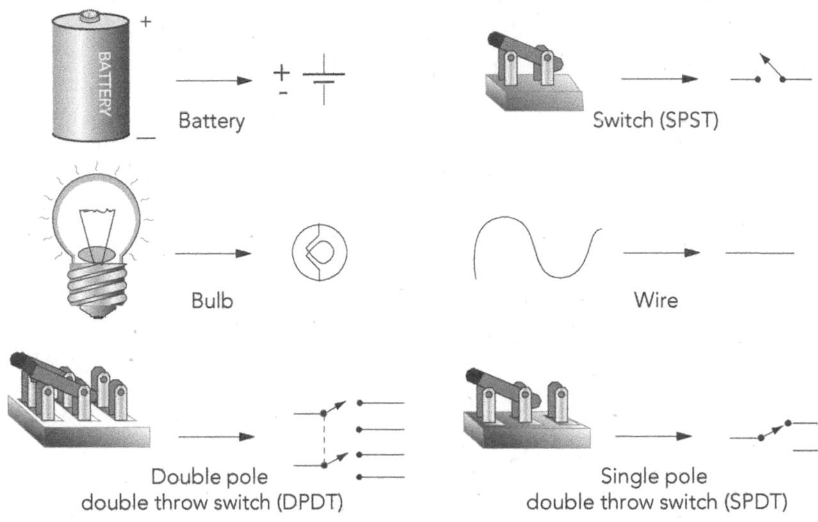

Fig. 22.8. Circuit symbols.

Using these symbols, the standard circuit with a switch, bulb, wires, and battery can be represented as shown in Fig. 22.9.

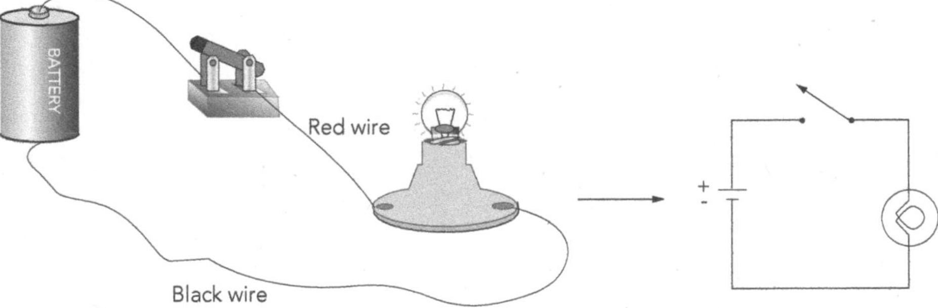

Fig. 22.9. A circuit sketch and corresponding diagram.

22.6.1. Activity: Drawing Circuit Diagrams

a. On the battery symbol, which line represents the positive terminal? The long one or the short one? **Note:** You should try to remember this convention for the battery polarity because some circuit elements, such as diodes, behave differently if the battery is turned around so it has opposite polarity.

b. Sketch nice neat "textbook"-style circuit diagrams for the circuits you designed in Activity 22.5.1.

CURRENT AND RESISTANCE

22.7. DEVELOPING A MODEL FOR CURRENT FLOW

Several models for current flow in the circuit have been proposed. Four are diagrammed below. Which diagram best describes your view of how current flows in the circuit? Why? Talk it over with your partners.

After you have discussed the various ideas with your instructor and the rest of the class, you will be asked to figure out how to use one or more ammeters in your circuit to measure current and test your model.

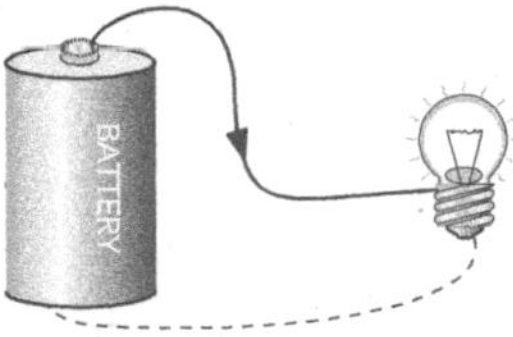

Model A
There will be no electric current left to flow in the bottom wire.

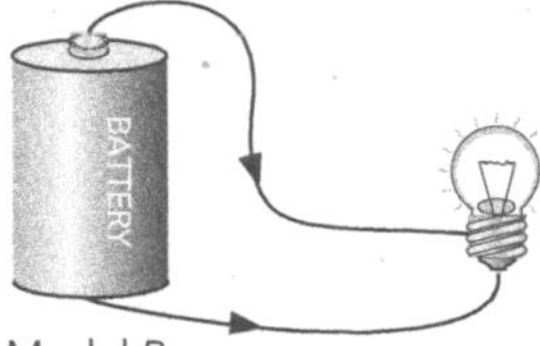

Model B
The electric current will travel in a direction toward the bulb in both wires.

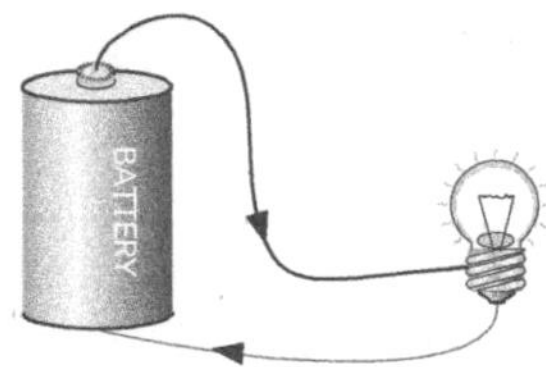

Model C
The direction of the current will be in the direction shown, but there will be less current in the return wire.

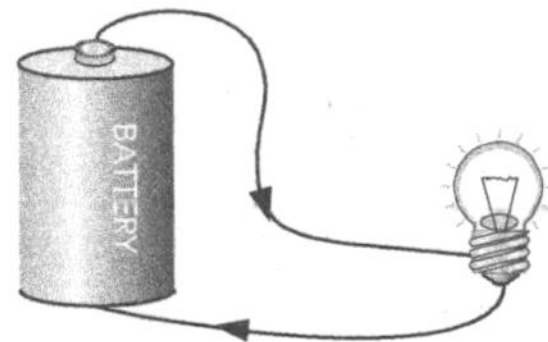

Model D
The direction of the current will be as shown, and it will be the same in both wires.

Fig. 22.10. Four alternative models for current flow.

22.7.1. Activity: Picking a Current Flow Model

Which model do you think best describes how current flows through the bulb? Explain your reasoning.

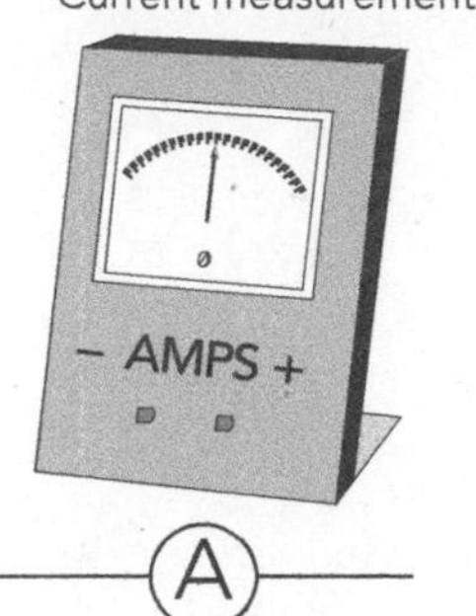

Fig. 22.11. An ammeter and its symbol.

22.8. MEASURING CURRENT WITH AN AMMETER

The ammeter is a device that measures current and displays it. You can use one or more ammeters to explore the current flowing at different locations in an electric circuit. To measure currents in a simple circuit containing a bulb you will need:

- 1 ammeter, 0.25 A
- 1 #14 bulb
- 1 #14 bulb holder
- 1 D-cell battery, 1.5 V
- 1 D-cell holder
- 1 SPST switch
- 4 alligator clip leads, > 10 cm

Recommended group size:	2	Interactive demo OK?:	N

Current is typically measured in amperes (A) or milliamperes (mA) with 1 ampere = 1000 milliamperes. Usually we just refer to current as "amps" or "milliamps."

To measure the current flowing *through* a part of a simple circuit with one battery and bulb like that shown in Figure 22.12, you must "insert" the ammeter at the point of interest. Disconnect the circuit, put in the ammeter, and reconnect it. For example, to measure the current in the bottom wire of the circuit in Figure 22.9, the ammeter could be connected as shown in Figure 22.12.

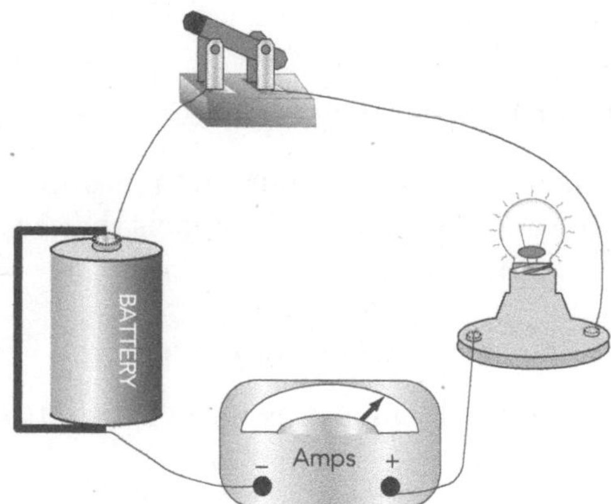

Fig. 22.12. Measuring current in a circuit with a battery, bulb, and ammeter.

22.8.1. Activity: Measuring Current

a. Set up the circuit shown in Figure 22.12. Measure the current through the ammeter. What is the value of the current in amps? Is it positive or negative?

b. Now reverse the leads going into and out of the ammeter. Explain what happens to the indicator needle.

22.9. DISCOVERING CURRENT FLOW MODELS THAT WORK

You can use one or more bulbs and one or more ammeters for this activity. You should have available:

- 2 ammeters, 0.25A
- 2 #14 bulbs (with identical brightness)
- 2 #14 bulb holders
- 1 D-cell battery, 1.5 V, alkaline
- 1 D-cell holder
- 1 SPST switch
- 6 alligator clip wires, > 10 cm

Discuss the various models with your partners and design measurements that will allow you to choose the model or models that best describe the actual current flowing through the circuit.

22.9.1. Activity: "Best" Current Flow Models

a. Describe your tests. Include drawings of the circuits you used, showing where the probes were connected.

b. Which model or models seem to work? Is the current different at different locations in the circuit in Figure 22.12? Explain how you reached your conclusion based on your observations.

22.10. MEASURING CURRENT AND POTENTIAL DIFFERENCE

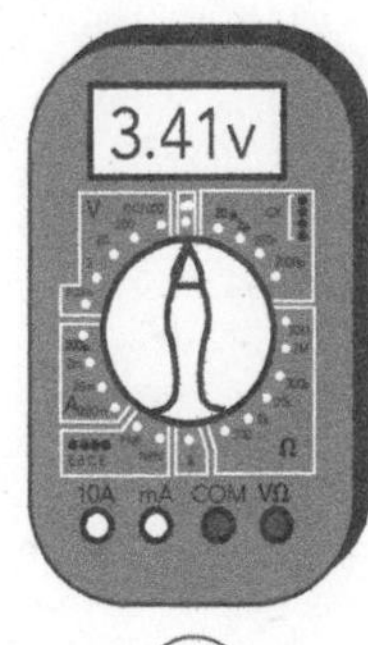

Fig. 22.13. A digital voltmeter and its symbol.

Since a battery is a device that has a potential difference *across* its terminals, it is capable of giving energy to charges, which can then flow as a current through a circuit. Exploring the relationship between the potential differences in a circuit and the currents that flow in that circuit is an essential part of developing an understanding of electrical circuits.

Since potential differences are measured in volts, a potential difference is informally referred to as a *voltage.*

Voltage is an informed term for potential difference. If you want to talk to physicists, you should refer to potential difference. Communicating with a sales person at the local Radio Shack store is another story. There you should probably refer to voltage. We will use these two terms interchangeably.

Let's measure voltage in a familiar circuit. Figures 22.14 and 22.15 show a simple circuit with a battery, a bulb, and two digital voltmeters connected to measure the voltage *across* the battery and the voltage *across* the bulb. The word *across* is descriptive of how the voltmeter connection is needed to measure voltage.

To do the next activity, you will need:

- 1 D-cell battery, 1.5 V, alkaline
- 1 D-cell holder
- 1 #14 bulb
- 1 #14 bulb holder
- 1 SPST switch
- 1 digital voltmeter

Note: If your digital voltmeter is actually a multimeter, be sure to set the dial for reading DCV (Direct Current Voltage).

Recommended group size:	2	Interactive demo OK?:	N

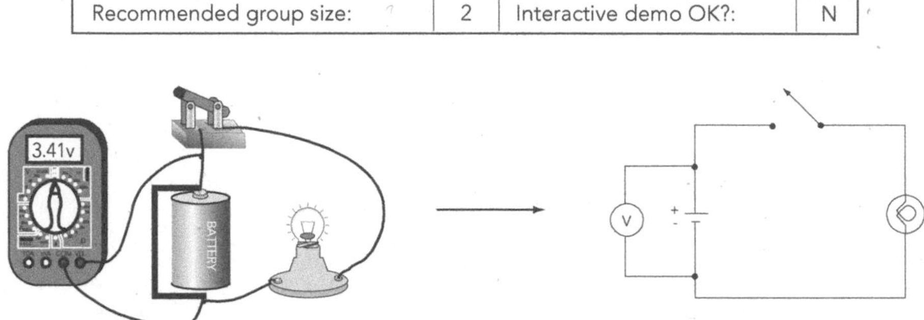

Fig. 22.14. Illustration of a circuit and a circuit diagram of a digital voltmeter used to measure the voltage across a battery in a simple circuit with a battery, a bulb, and a switch.

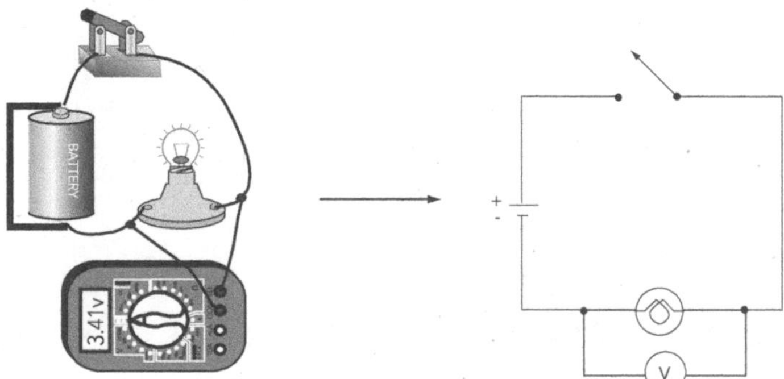

Fig. 22.15. Illustration of a circuit and a circuit diagram of a digital voltmeter used to measure the voltage across a bulb in a simple circuit with a battery, a bulb, and a switch.

22.10.1. Activity: Voltage Measurements

a. First connect both the positive and negative leads of the voltmeter to the *same point* in the circuit. Observe the reading. Repeat this process of measuring voltage at another point in the circuit. What do you conclude about the voltage when the leads are connected to each other (that is, are not across anything else)?

b. In the circuit in Figure 22.14, predict the voltage across the battery compared to the voltage across the bulb. Explain your predictions.

c. Test your prediction by connecting the circuit in Figure 22.14. Use the digital voltmeter to measure the voltage *across* the battery and then use it to measure the voltage *across* the bulb. How are the voltage across the battery and the voltage across the bulb related?

22.11. CURRENT, VOLTAGE, AND MORE BULBS

Now that you know how to measure both current through a circuit and the potential difference (voltage) between or across two points in a circuit you can begin to explore the relationship between these two quantities as we add more bulbs to a circuit. You can start by measuring the current through a circuit and the potential difference across a battery with only one bulb in it. What do you think will happen to the voltage across the battery when a second bulb is added to the circuit? What will happen to the current in the circuit?

To do the next few activities, you will need:

- 1 D-cell battery, 1.5 V, alkaline
- 1 D-cell holder
- 2 #14 bulbs (with the same brightness)
- 2 #14 bulb holders
- 1 SPST switch
- 1 digital voltmeter
- 1 ammeter, 0.25 A

Recommended group size:	2	Interactive demo OK?:	N

Now let's measure voltage across the battery and current in your single bulb circuit at the same time. To do this, connect your voltmeter and an ammeter so that you are measuring the voltage across the battery and the current through the circuit at the same time. (See Figure 22.16.) **Note:** Before proceeding with the next activity, set up the circuit in Fig. 22.1b and test each bulb in the circuit to confirm that your bulbs have essentially the same brightness.

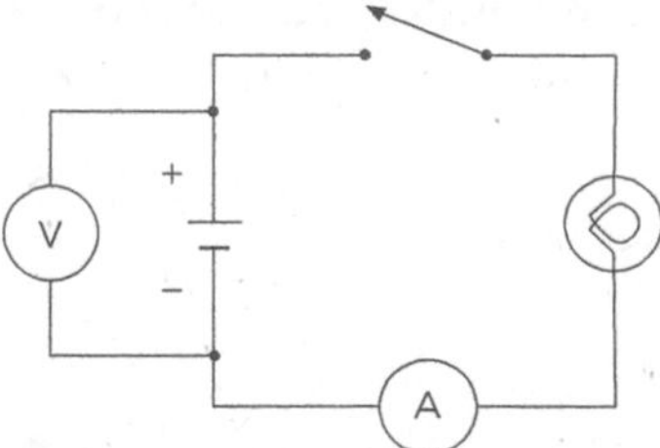

Fig. 22.16. A circuit diagram showing connections to measure the voltage across a battery and current through the circuit.

22.11.1. Activity: Single Bulb Current/ Voltage Measurements

a. Measure the voltage across the battery when the switch is closed and the light is lit.

Voltage across the battery (with one bulb): ________ Volts

b. Measure the current through the circuit, that is, drawn from the battery when the switch is closed and the light is lit.

Current drawn from the battery (with one bulb): ________ Amps

Now suppose you connect a second bulb in the circuit as shown in Figure 22.17.

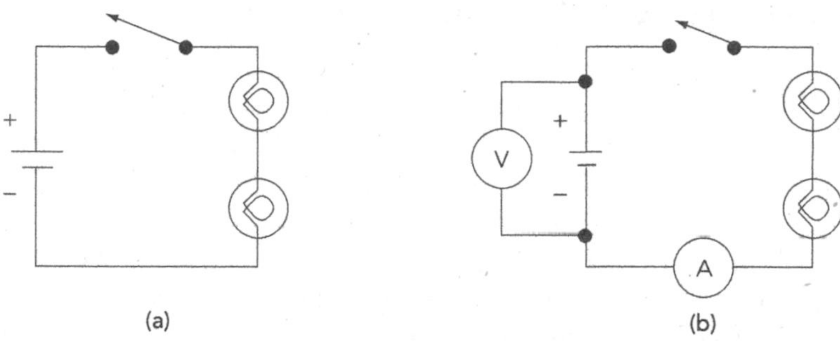

Fig. 22.17. Two bulbs connected to a battery without meters attached and with a voltmeter and ammeter set up to measure voltage across the battery and through the circuit.

22.11.2. Activity: Current and Voltage for Two Bulbs

a. How do you think the voltage across the battery will compare to that with only one bulb? Will it change significantly?

b. What do you think will happen to the brightness of the first bulb when you add a second bulb to the circuit? Will it get brighter? Dimmer? Remain the same?

c. What will happen to the current drawn from the battery? Will it remain the same, decrease, or increase?

d. Explain the reasons for your predictions in parts a., b., and c. of this activity.

e. Add a second bulb to the circuit and test your predictions. Again measure the voltage across the battery with the switch closed. What is the voltage?

Voltage across the battery (with two bulbs): __________ Volts

f. Did the voltage across the battery change significantly? Did the addition of the second bulb seem to change the potential difference across the battery?

g. Did the first bulb dim when you added the second one to the circuit? What happens to the current drawn from the battery?

Current drawn from the battery (with two bulbs): __________ Amps

h. Explain what you think the battery is doing and why the differences in voltage and current (if any) occur when a second bulb is added to the circuit.

i. Is the battery a source of constant current? Why or why not?

22.12. AN ANALOGY TO POTENTIAL DIFFERENCE AND CURRENT FLOW

The fact that you found that a current is not "used up" in passing through a bulb seems counterintuitive to many people trying to understand how circuits work. Many physics teachers have invented analogies to help explain this idea for an electric circuit. One obvious approach is to construct a model of a gravitational system that is in some ways analogous to the electrical system we are studying.

It is believed that the electrons flowing through a conductor collide with the atoms in the material and scatter off them. After colliding with an atom, each electron accelerates again until it collides with another atom. In this manner, the electron finally staggers through the material with an average drift velocity $<\vec{v}>$.

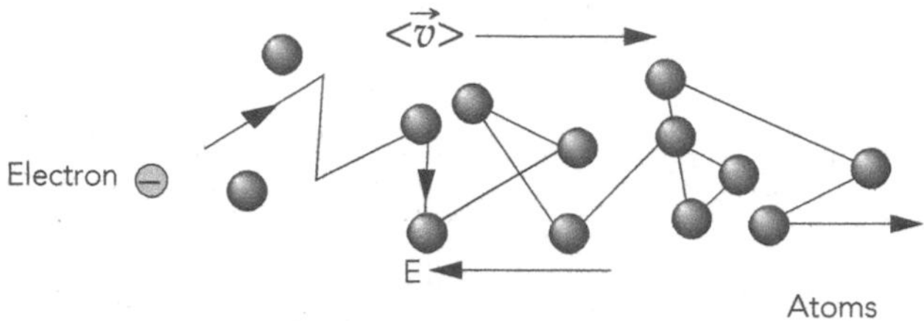

Fig. 22.18. An electron in a uniform electric field staggering through a conductor as a result of collisions with atoms. Instead of accelerating it has an average drift velocity, $<\vec{v}>$.

We can talk about the *resistance* to flow of electrons that materials offer. A thick wire has a low resistance. A light bulb with a thin filament has a much higher resistance. Special electric elements that resist current flow are called resistors. We will examine the behavior of *resistors* in circuits in future activities.

It is possible to use a two-dimensional mechanical analog to model this picture of current flow through conductors. You should note that the real flow of electrons is a three-dimensional affair. The diagram for the two-dimensional analog is reproduced in Figure 22.19.

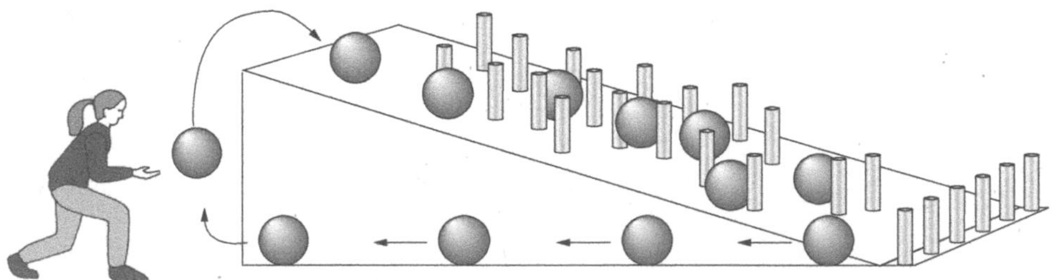

Fig. 22.19. An analog to electrical current flow.

Building a Working Model for Current Flow

In order to see how the model works, you should build it and play with it to see how it works.

To build the model you can take an adjustable wooden ramp and mount a piece of insulation board on it. Then push pins can be poked into the soft insulation board to simulate atoms. Place graph paper on top of the insulation board to facilitate the spacing of the atoms. A marble can be used to represent an electron flowing through the circuit. Strips of poster board can be pinned into the sides of the insulation board to make side rails. For this project you will need the following items:

- 6 identical steel balls
- 1 ramp (with adjustable angles)
- 1 ruler
- 1 protractor
- 1 board with face-centered pegs or
    - 1 polyurethane insulation board, approx. 6" × 18"
    - 2 poster boards, approx. 2" × 18"
    - 100 plastic push pins
    - 2 graph paper sheets, 8.5" × 11" with 1/4" grid lines
- 1 digital stopwatch

Recommended group size:	3	Interactive demo OK?:	N

Building a Model

If you need to build a model, the push pins should represent a face-centered array of atoms. A three-dimensional face-centered arrangement of atoms is shown in the diagram below. How could you represent this in two dimensions? The maximum spacing between corner atoms should be large enough to allow balls to "flow," but small enough to force regular collisions. Don't forget to put an atom in the center of each set of four corner atoms.

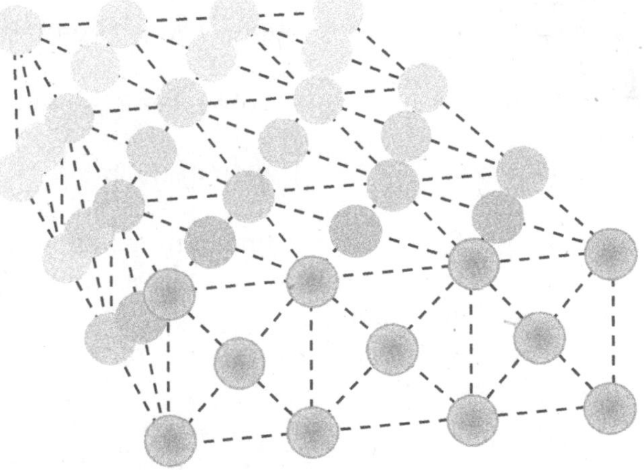

Fig. 22.20. Arrangement of atoms in a face-centered cubic crystal. The atoms in the foreground are represented by dark spheres.

By placing the board with face-centered pegs on the ramp and tilting it at an angle you can investigate the analogy between gravitational potential energy and the electrical potential energy stored in a battery. Suppose that the ramp is propped up so that one end is a height h above the other end and that a ball is rolled down the ramp with the pegs providing resistance to the flow of balls, and that a little person keeps lifting up a ball to the top of the ramp as soon as it reaches the bottom.

22.12.1. Activity: Explaining the Features of the Model

a. What would happen to the ball current (that is, the rate of ball flow) if twice as many pegs were placed in the path of the ball?

b. What would happen to the ball current if the ramp were raised to a height of $2h$ so that balls have twice as much gravitational potential energy when they start to fall?

c. Examine the list below and *draw lines between elements of the model and the corresponding elements of a circuit consisting of batteries and bulbs* (or other electrical resistors). In particular, what represents the electrical charge and current? What ultimately happens to the "energy" given to the bowling balls by the "battery"? What plays the role of the bulb? Where might mechanical energy loss occur in the circuit you just wired that consists of a battery, two wires, and a bulb?

Electrical	Mechanical Analogy
Number of bulbs (resistance)	Average speed of bowling balls
Battery action	Number of pegs encountered
Current	Person raising the balls
Voltage of battery	Height of the ramp

d. What ultimately happens to the gravitational potential energy given to the balls by the person lifting the balls as the balls fall down the ramp? What do you think might happen to the electrical potential energy of an electron "lifted" by the action of the battery as it passes through a bulb?

e. How can this model be used to explain why electric current isn't used up when it flows through a bulb?

f. How does this ramp analogy support a model in which current doesn't accelerate when it flows through a circuit?

g. In this model, what would happen to the "ball" current if the drift velocity doubles? What can you do to the ramp to increase the drift velocity?

SERIES AND PARALLEL CIRCUITS

22.13. OVERVIEW

In the last several activities you saw that, when an electric current flows through a light bulb, the bulb lights. You also saw that to get a current to flow through a bulb you must connect the bulb in a complete circuit with a battery. A current will only flow when it has a complete path from the positive terminal of the battery, through the connecting wire to the bulb, through the bulb, through the connecting wire to the negative terminal of the battery, and through the battery.

By measuring the current at different points in the simple circuit consisting of a bulb, a battery, and connecting wires, you have discovered a model for current flow—namely, that the electric current was the same in all parts of a simple circuit. By measuring the current and voltage in this circuit and adding a second bulb in series with the first one, you also should have discovered that a battery supplies essentially the same voltage whether it is connected to one light bulb or two. You also discovered that less current is drawn from the battery when there are two bulbs in series. Two bulbs seem to offer more "resistance" to the flow of current than one bulb does.

In the next several activities you will examine more complicated circuits than a single bulb connected to a single battery. You will compare the currents through different parts of these circuits in which elements are placed both in series and in parallel with each other. You will do this by comparing the brightness of the bulbs, and also by measuring currents with ammeters.

22.14. DEVISING RULES TO EXPLAIN CURRENT FLOW

In the next series of exercises you will be asked to make a number of predictions and then to confirm these predictions with observations. Whenever your observations and predictions disagree, you should try to develop new concepts about how circuits with batteries and bulbs actually work. In order to make the required observations you should wire circuits with a *very fresh* alkaline battery. You'll need:

- 1 D-cell battery, 1.5 V, alkaline
- 1 D-cell holder
- 6 alligator clip leads, > 10 cm
- 4 #14 bulbs (with identical brightness)
- 4 #14 bulb holders
- 1 SPST switch
- 1 ammeter, 0.25 A

Recommended group size:	2	Interactive demo OK?:	N

Let's start with a quick review of observations made in the last few activities. Consider the two circuits shown in Figure 22.21. Check to see that all your bulbs have the same brightness. You will be asked to predict the relative brightness of the various bulbs. Recall that a very fresh alkaline battery supplies essentially the same voltage whether there is one light bulb or two.

> **Hint:** Helpful symbols for reporting relative brightnesses are
>
> > "greater than"
> < "less than"
> = "equal to."

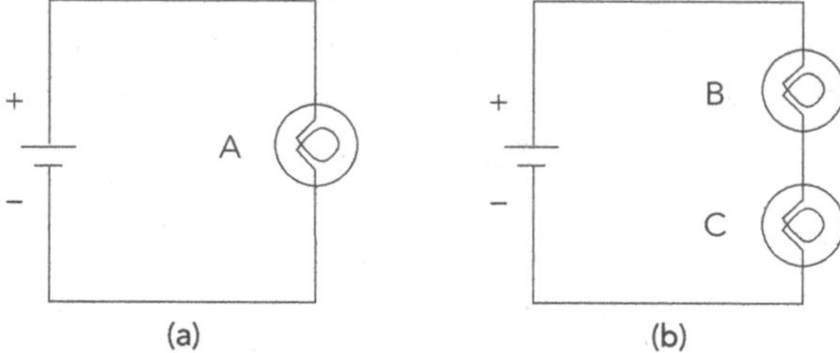

Fig. 22.21. Two different circuits with identical components: a battery with a single bulb and an identical battery with two bulbs.

22.14.1. Activity: Relative Brightnesses of Bulbs

a. Before doing any observations, predict the relative brightness of the three bulbs shown in Figure 22.21, from brightest to dimmest. If two or more bulbs are equal in brightness, indicate this in your response. *Explain the reasons for your rankings.*

b. Now connect the circuits, observe the relative brightnesses of the bulbs, and rank the bulbs in order of the brightnesses you actually observed.

c. Did your observations agree with your predictions? If not, can you explain what assumptions you were making that seem false?

d. If we think of the bulb as providing a resistance to the flow of current in a circuit, rather than as something that uses up current, how do you think the total resistance to the flow of current through a circuit will be affected by the addition of more bulbs in the manner shown in Figure 22.21b? **Note:** You may want to review Activities 22.9 and 22.11.

e. Formulate a rule for predicting whether current increases or decreases as the total resistance of the circuit is increased.

Note: The rule you probably formulated based on your observations with bulbs may be *qualitatively* correct—correctly predicting an increase or decrease in current—but it won't be *quantitatively* correct. That is, it won't allow you to predict the exact sizes of the currents correctly. This is because the resistance of a bulb to current flow changes as the current through the bulb changes. Later we will find a more quantitative law using circuit elements called *resistors* that have a constant resistance regardless of the current.

22.15. CURRENT AND VOLTAGE IN PARALLEL CIRCUITS

There are two basic ways to connect resistors (such as bulbs) in a circuit—*series* and *parallel.* So far you have been dealing with bulbs wired in *series.* To make predictions involving more complicated circuits we need to have more precise definitions of the terms *series* and *parallel.* These are summarized in Figure 22.22.

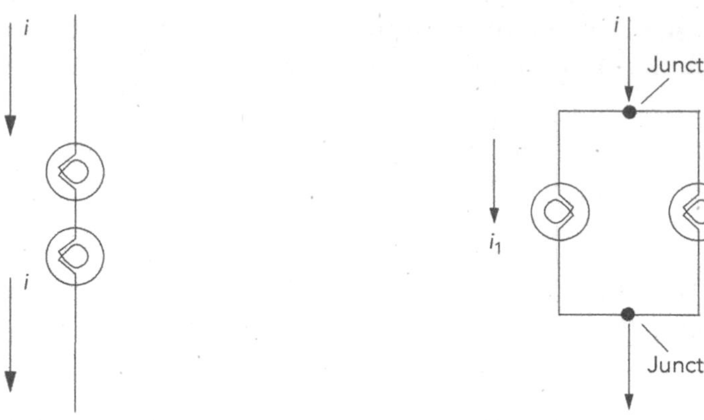

Series connection:
Two resistors are in series if they are connected so that the same current that passes through one bulb passes through the other.

Parallel connection:
Two resistors are in parallel if their terminals are connected so that at each junction one terminal of one bulb is directly connected to one terminal of the other.

Fig. 22.22. (a) Series connection and (b) parallel connection.

Let's compare the behavior of a circuit with two bulbs wired in parallel to the circuit with a single bulb. (See Figure 22.23.)

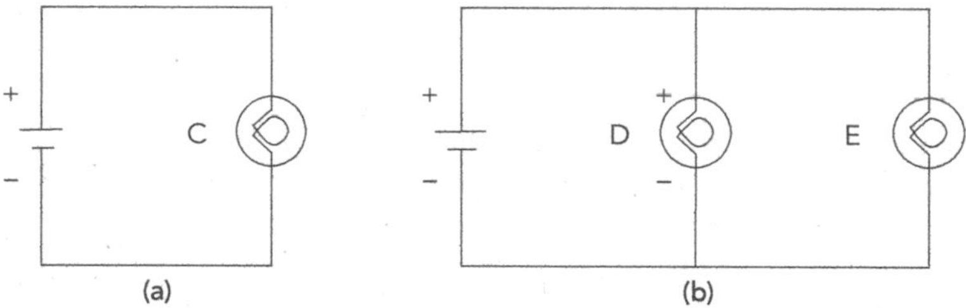

Fig. 22.23. Two different circuits with identical components: (a) a single bulb circuit and (b) a parallel circuit.

Note that if bulbs C, D, and E are identical, then the circuit in Figure 22.24 is equivalent to the circuit in Figure 22.23a when the switch is open (as shown) and equivalent to the circuit in Figure 22.23b when the switch is closed.

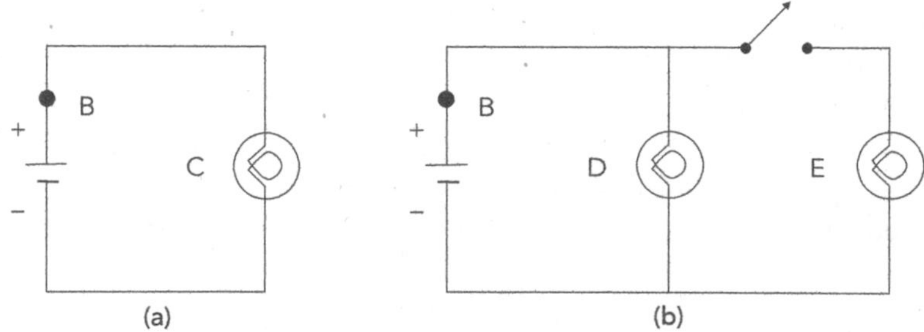

Fig. 22.24. When the switch is open, only bulb D is connected to the battery. When the switch is closed, bulbs D and E are connected to the battery in parallel.

To make the predictions and observations in the next two activities you will need a fresh alkaline battery and other circuit elements:

- 1 D-cell battery, 1.5 V, alkaline (fresh)
- 1 D-cell holder
- 3 #14 bulbs (with identical brightness)
- 3 #14 bulb holders
- 6 alligator clip leads, > 10 cm
- 2 ammeters, 0.25 A
- 1 digital voltmeter
- 1 SPST switch

Recommended group size:	2	Interactive demo OK?:	N

Let's start by considering the current in the branches of a parallel circuit.

22.15.1. Activity: Predicting Relative Currents in a Parallel Circuit

a. Predict the relative brightness of bulb C in the circuit in Figure 22.23a with the brightnesses of bulbs D and E in Figure 22.23b. Which of the three bulbs will be the brightest? The dimmest? Will any of the bulbs be the same brightness? Explain the reasons for your prediction. Don't forget to use the following symbols for reporting predictions and observations of relative brightnesses: > "greater than", < "less than", = "equal to".

b. How do you think that closing the switch in Figure 22.24 will affect the current through bulb D? Explain!

You can test predictions you just made in Activity 22.15.1 by wiring up the circuit shown in Figure 22.24 and looking at what happens to the brightness of the bulbs as the switch is closed.

Note: Before you start the next activity, make sure that (1) bulbs D and E have the *same* brightness when placed in series with the battery and (2) use a very fresh alkaline D-cell battery so it behaves like an "ideal" battery.

22.15.2. Activity: Actual Currents in the Parallel Circuit

a. Wire up the circuit in Figure 22.24 and, by opening and closing the switch, describe what you observe to be the actual relative brightness of bulbs C, D, and E. **Note:** Your battery may not be *ideal,* and you should not report a very small change in bulb brightness as a "change."

b. Did closing the switch and connecting bulb E in parallel with bulb D significantly affect the current through bulb D? How do you know?

c. Based on your observations, were the relative currents through bulbs C, D, and E what you predicted? If not, can you now see why your prediction was incorrect?

What About the Battery Voltages and Current?

Is the battery current always the same no matter what is connected to it, or does it change depending on the circuit? In other words, is the current through the battery the same whether the switch in Figure 22.24 is open or closed? Does the voltage across the battery change when the switch is closed? Make predictions based on your observations of the relative brightness of the bulbs, and test them using meters to measure voltage and currents.

22.15.3. Activity: Predicting Changes in Battery Current and Voltage

a. Based on your observations in Activity 22.15.2, how do you think that closing the switch in Figure 22.24 will affect the current *through* the battery—that is, *through* point B? Explain the reasons for your prediction.

b. Based on your observations in Activity 22.15.2, how do you think that closing the switch in Figure 22.24 will affect the current *through* bulb D? Explain the reasons for your prediction.

c. Based on your observations in Activity 22.15.2, how do you think that closing the switch in Figure 22.24 will affect the voltage *across* the battery? Explain the reasons for your prediction.

Test your predictions by placing the voltmeter across the battery and inserting the ammeters in the circuit as shown in the following circuit diagram.

Note: Before you start the next activity, make sure that (1) bulbs D and E have the *same* brightness when placed in series with the battery and (2) use a very fresh alkaline D-cell battery so it behaves like an "ideal" battery.

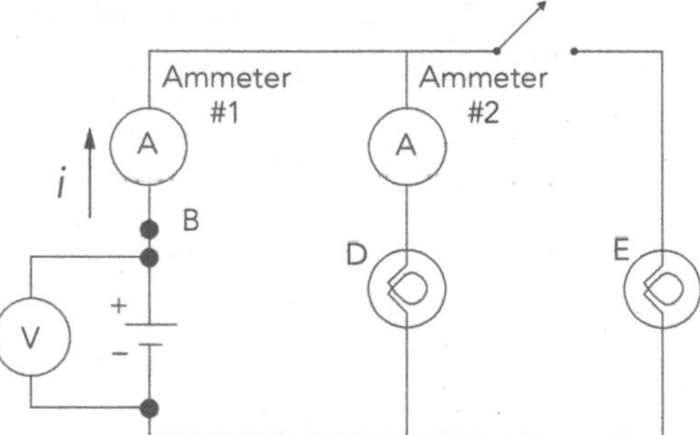

Fig. 22.25. Meters connected to measure the current through bulb D and the current through point B and the battery and the voltage across the battery when the switch is opened and closed.

22.15.4. Activity: Observing Battery Voltage and Current

a. How does closing the switch in a circuit like that shown in Figure 22.25 affect the current through the battery—that is, *through* point B?

Battery current before the switch is closed: ________Amps

Battery current after the switch is closed: ________Amps

b. How does closing the switch in a circuit like that shown in Figure 22.25 affect the current through bulb D?

Bulb D current before the switch is closed: ________Amps

Bulb D current after the switch is closed: ________Amps

c. How does closing the switch in a circuit like that shown in Figure 22.25 affect the voltage across the battery? Explain the reasons for your prediction.

Battery voltage before the switch is closed: __________ Volts

Battery voltage after the switch is closed: __________ Volts

d. Use your observations to formulate a rule to predict how the current through a battery will change as the number of bulbs connected in *parallel* increases. Can you explain why?

e. Compare your rule in d. to the rule you stated in Activity 22.14.1e relating the current through the battery to the total *resistance* of the circuit connected to the battery. Does the addition of more bulbs in parallel increase, decrease, or not change the total *resistance* of the circuit?

f. Explain your answer to d. in terms of the number of paths available in the circuit for current flow.

g. Does the amount of current through a battery appear to depend only on how many bulbs are in the circuit or does the arrangement of the bulbs matter also? (Don't forget your observations with bulbs in series in Activity 22.11.2g and with bulbs in parallel in Activity 22.14.4a.)

h. Does the total resistance of a circuit appear to depend only on how many bulbs are in the circuit or does the arrangement of the bulbs matter also?

22.16. MORE COMPLEX SERIES AND PARALLEL CIRCUITS

To test your predictions and make observations associated with the activities in this section you will need a fresh alkaline battery and other circuit elements:

- 3 D-cell batteries, 1.5 V, alkaline (fresh)
- 3 D-cell holder

OR

- 1 6 V lantern battery

- 3 #14 bulbs (with identical brightness)
- 3 #14 bulb holders
- 1 SPST switch
- 2 ammeters, 0.25 A
- 1 digital voltmeter

Recommended group size:	2	Interactive demo OK?:	N

Applying Your Knowledge to a More Complex Circuit

Consider the circuit consisting of a battery and two bulbs, B and C, in series shown in Figure 22.26a. What will happen if you add a third bulb, D, in parallel with bulb C (as shown in Figure 22.26b)? You should be able to answer this question about the relative brightness of B, C, and D based on previous observations. The tough question is: how does the brightness of B change?

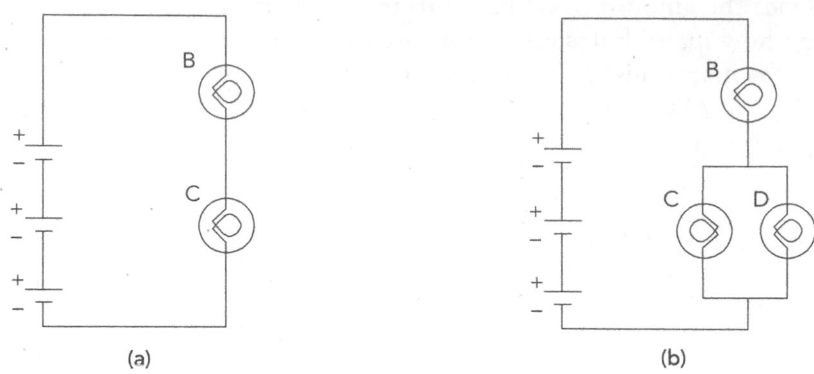

Fig. 22.26. Two different circuits with identical bulbs.

22.16.1. Activity: Predictions for a Complex Circuit

a. In Figure 22.26b, are C and D in series or in parallel with each other? Explain.

b. In Figure 22.26b, is B in series with C alone, with D alone, or with a parallel combination of C and D? (You may want to review the definitions of series and parallel connections given earlier in this unit.)

c. Is the resistance of the combination of C and D larger than, smaller than, or the same as B alone? Explain.

d. Is the resistance of the combination of bulbs B, C and D in Figure 22.26b larger than, smaller than, or the same as the combination of bulbs B and C in Figure 22.26a? Explain.

e. Can you predict how the current through bulb B will change, if at all, when the circuit in Figure 22.26a is changed to the one shown in Figure 22.26b (that is, when bulb D is added in parallel to bulb C)? Explain the reasons for your answer.

f. When bulb D is added to the circuit, what will happen to the brightness of bulb B? Explain.

g. Finally, predict the relative rankings of brightness for all the bulbs, B, C, and D, after bulb D has been added to the circuit. Explain the reasons for your answers.

Convince yourself that the circuit in Figure 22.27a is identical to Figure 22.26a when the switch is open (as shown) and to 22.26b when the switch is closed and then make some observations using the circuits in Figure 22.27.

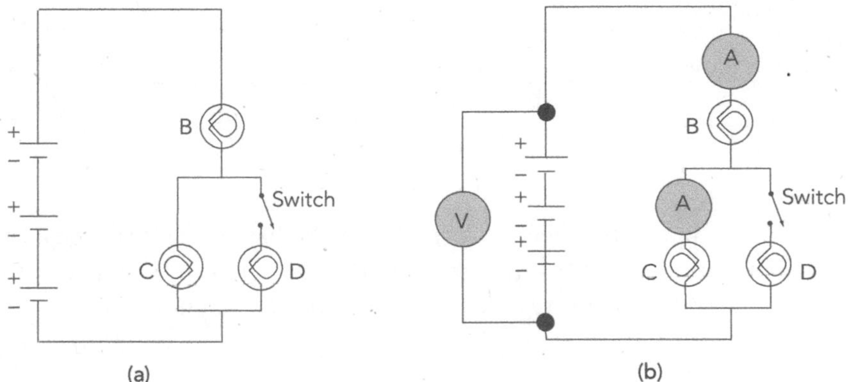

Fig. 22.27. (a) Circuit equivalent to Figure 22.26a when the switch is open, and to Figure 22.26b when the switch is closed. (b) Same circuit with ammeters and a voltmeter connected to measure the current through bulb B, the current through bulb C and the voltage across the battery.

22.16.2. Activity: Observing a Complex Circuit

a. Set up a circuit like that in Figure 22.27a and observe the brightness of bulbs B and C when the switch is open, and then the brightness of the three bulbs when the switch is closed (when bulb D is added). Compare the brightness of bulb B with and without D added, and rank the brightness of bulbs B, C, and D with the switch closed.

b. If your observations and predictions of the brightnesses were not consistent, what changes do you need to make in your reasoning? Explain!

c. Connect the voltmeter and the two ammeters as shown in Figure 22.27b. Record what happens to the voltage across the battery combination, the current through bulb B, and the current through bulb C when the switch is opened and closed. Record your measurements.

Measurement	Before switch closed	After switch closed
Combined battery voltage	Volts	Volts
Bulb B current	Amps	Amps
Bulb C current	Amps	Amps

d. What happens to the current through the battery and through bulb C when bulb D is added in parallel with bulb C? What do you conclude happens to the total resistance in the circuit? Explain.

Series and Parallel Networks

Let's look at a somewhat more complicated circuit to see how series and parallel parts of a complex circuit affect one another. The circuit is shown in Figure 22.28.

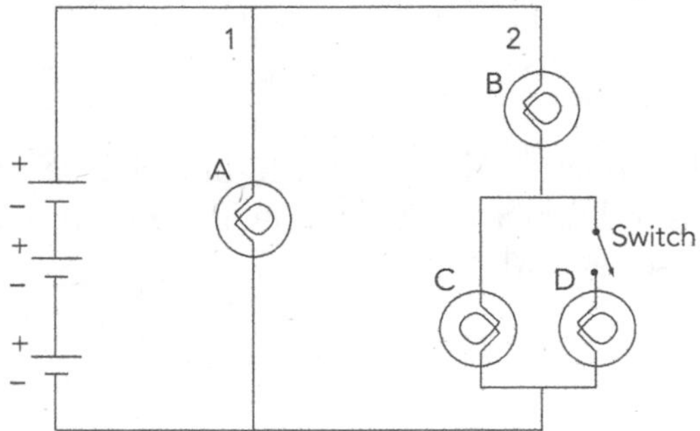

Fig. 22.28. A complex circuit with series and parallel connections.

22.16.3. Activity: Series and Parallel Network

a. When the switch is open, which bulbs are connected in parallel with each other? Is A parallel to B? Is A parallel to C? Is C parallel to D? Is A parallel to the combination of B and C? If you need to, review the definitions of series and parallel connections given earlier in this unit before answering.

b. When the switch is open, which bulbs are connected in series with each other? Is A in series with B? Is A in series with C? Is B in series with C?

c. When the switch is closed, which bulb(s) are connected in parallel with A?

d. When the switch is closed, which bulb(s) are connected in series with B?

e. Predict the effect on the current in branch 1 of each of the following alterations in branch 2: (1) unscrewing bulb B, and (2) closing the switch.

f. Predict the effect on the current in branch 2 of each of the following alterations in branch 1: (1) unscrewing bulb A, and (2) adding another bulb in series with bulb A.

g. Connect the circuit in Figure 22.28, and observe the effect of each of the alterations in e. and f. on the brightness of each bulb. *Record your observations for each case.*

h. Compare your results with your predictions. How do you account for any differences between your predictions and observations?

i. In this circuit, two parallel branches are connected across a battery. What do you conclude about the effect of changes in one parallel branch on the current in the other?

Name ______________________ Section ____________ Date ____________

UNIT 23: DIRECT CURRENT CIRCUITS

As a result of atmospheric processes, negative charges build up at the base of clouds causing a potential difference between them and the ground beneath the cloud. During lightning storms, this charge difference is dissipated by the flow of electrons downward and positive ions upward along a complex network of paths. If you knew the effective resistance of each of the discharge paths shown above, and the potential difference between the cloud and the ground, how would you calculate the current in each path? In this unit you will learn how to use the principles of charge and energy conservation to develop a system for determining currents for each path in a complex circuit.

UNIT 23: DIRECT CURRENT CIRCUITS*

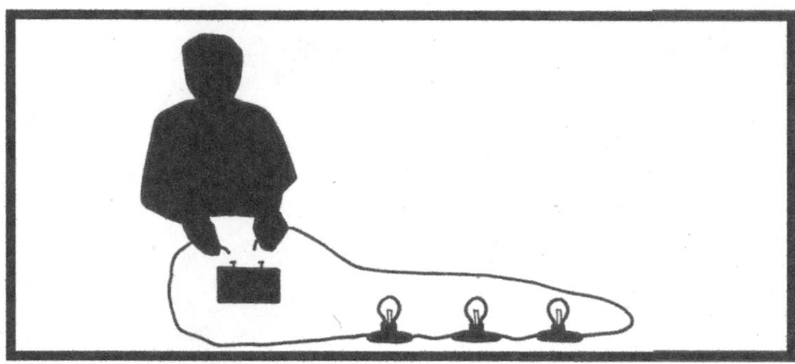

I have a strong resistance to understanding the relationship between voltage and current. Anonymous introductory physics student

OBJECTIVES

1. To learn to apply the concept of potential difference or voltage to explain the action of a battery in circuits.

2. To understand the distribution of potential difference or voltage in all parts of a series circuit.

3. To understand the distribution of potential difference or voltage in all parts of a parallel circuit.

4. To understand and apply the relationship between potential difference and current for a resistor with negligible temperature dependence (Ohm's law).

5. To find a mathematical description of the flow of electric current through different elements in direct current circuits (Kirchhoff's laws).

6. To gain experience with basic electronic equipment and the process of constructing useful circuits while reviewing the application of Kirchhoff's laws.

* Portions of this unit are based on research by Lillian C. McDermott and Peter S. Shaffer published in *AJP 60*, 994–1012 (1992).

23.1. OVERVIEW

In the last two units you explored currents at different points in series and parallel circuits. You saw that in a series circuit, *the current is the same through all elements.* You also saw that in a parallel circuit, *the current divides among the branches so that the current through the battery equals the sum of the currents in each branch.*

You have also observed that making a change in one branch of a parallel circuit does not affect the current flowing in the other branch (or branches), while *changing one part of a series circuit changes the current in all parts of the circuit.*

In observing series and parallel circuits, you have seen that certain connections of light bulbs result in a larger resistance to current flow while others result in a smaller resistance to current flow.

In this unit, you will first examine the role of the battery in causing a current to flow in a circuit. You will then compare the potential differences (voltages) across different parts of series and parallel circuits.

In the next few activities you will explore the relationship between the current through a resistor and the potential difference (voltage) across the resistor; this relationship is known as Ohm's law.

In addition, you will measure the effective resistance of carbon resistors when they are wired in series and in parallel. Finally, you will formulate the rules for the calculation of the electric current in different parts of complex electric circuits consisting of many resistors and/or batteries wired in series and parallel. These rules are known as Kirchhoff's laws. To test your understanding of Kirchhoff's laws, you will learn to use a protoboard to wire complex electric circuits. By measuring the current in different parts of your circuit you can verify that your theoretical application actually describes "reality."

BATTERIES AND POTENTIAL DIFFERENCES IN SERIES AND PARALLEL CIRCUITS

23.2. POTENTIAL DIFFERENCE WHEN BATTERIES ARE COMBINED

In the following activities you will continue to explore voltages and currents in circuits when batteries are combined.

In order to do this you will need the following items:

- 2 D-cell batteries, 1.5 V, alkaline (fresh)
- 2 D-cell holders
- 6 alligator clip leads, > 10 cm
- 4 #14 bulbs (with identical brightness)
- 4 #14 bulb holders
- 1 SPST switch
- 1 digital voltmeter
- 1 ammeter, 0.25 A

Recommended group size:	2	Interactive demo OK?:	N

You have already seen what happens to the brightness of the bulb in the circuit in Figure 23.1a if you add a second bulb in series as in the circuit in Figure 23.1b. The two bulbs are less bright than the original bulb because their resistance is larger, resulting in less current flow through the bulbs.

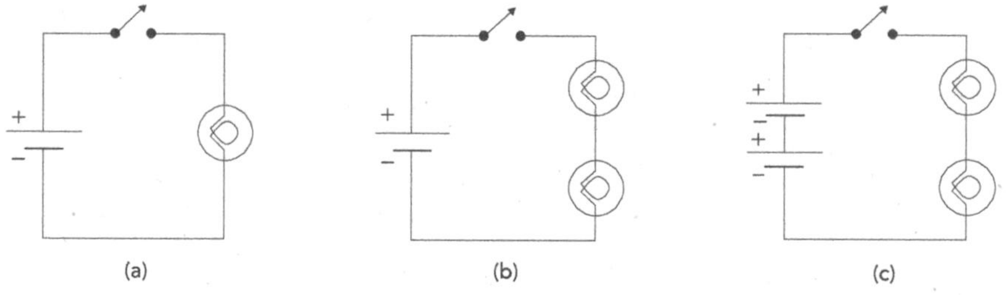

Fig. 23.1. Series circuits with one battery and one bulb; one battery and two bulbs; and two batteries and two bulbs.

Let's make some predictions about the result of adding a second bulb and battery to a single bulb single battery circuit.

23.2.1. Activity: Adding a Second Battery and Bulb in Series

a. What do you predict would happen to the brightness of the bulbs in Figure 23.1 if you connected a second battery in series with the first at the same time you added the second bulb (as in Figure 23.1c)?

b. Connect the circuit in Figure 23.1a. Then connect the circuit in Figure 23.1c. Compare the brightness of each of the bulbs in Figure 23.1c to the brightness of the single bulb in Figure 23.1a.

c. What do you conclude about the current in the two-bulb, two-battery circuit as compared to the single-bulb, single-battery circuit?

d. What happens to the resistance of a circuit as more bulbs are added in series? What must you do to keep the current from decreasing?

Let's compare the brightness of the bulb in the circuit below (Figure 23.2) to the brightness of the bulb in Figure 23.1a.

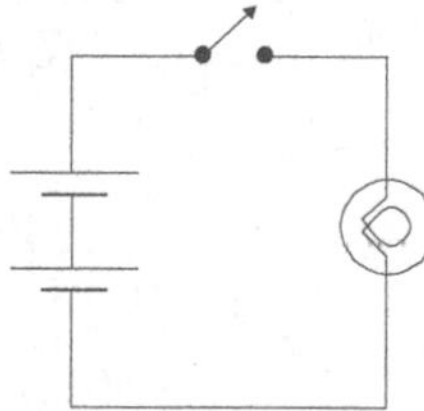

Fig. 23.2. Series circuit with two batteries and one bulb.

23.2.2. Activity: Adding a Second Battery

a. What do you predict will happen to the brightness of the bulb if a second battery is added? Explain the reasons for your prediction.

b. Connect the circuit in Figure 23.2. *Only close the switch for a moment to observe the brightness of the bulb—otherwise, you might burn out the bulb.* Compare the brightness of the bulb to the single bulb circuit with only one battery (Figure 23.1a).

c. How does increasing the number of batteries connected in series affect the current in a series circuit?

d. What characteristic of the battery determines the bulb brightnesses?

Fig. 23.3.

Let's explore potential differences when batteries are wired in series and parallel circuits, and see if you can develop rules to describe its behavior as we did earlier for currents. Figure 23.4 shows a single battery, two batteries identical to it connected in series, and then two batteries identical to it connected in parallel.

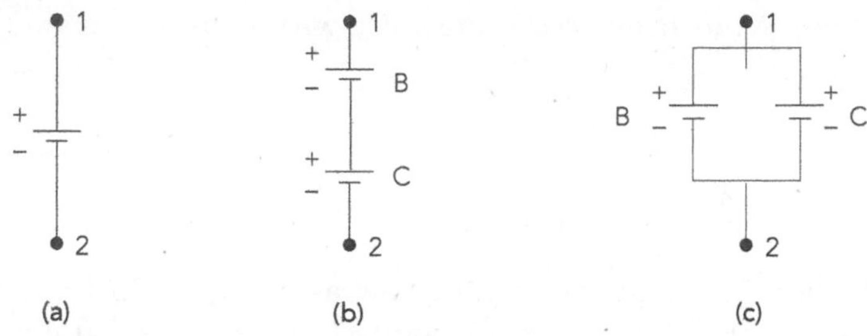

Fig. 23.4. A single battery; two identical batteries connected in series; and two connected in parallel.

You can measure potential differences with the digital voltmeter.

23.2.3. Activity: Batteries in Series and Parallel

a. If the potential difference between points 1 and 2 in Figure 23.4a is known to be V, predict the potential difference between points 1 and 2 in the series connection shown in Figure 23.4b and in the parallel connection shown in Figure 23.4c. Explain your reasoning.

b. Measure the size of the potential differences of battery B and of battery C. Record the voltage measured for each battery below.

Size of Potential Difference Across:

Battery B: ________

Battery C: ________

c. How do your measured values agree with those marked on the batteries?

d. Now connect the batteries in series (as in Figure 23.4b) and measure the potential difference across points 1 and 2 to determine the potential difference across the series combination of the two batteries. Record your measured value.

Potential difference across B and C in series: ________ Volts

e. How does your measured value agree with your prediction?

f. Now connect the batteries in parallel as in Figure 23.4c, and measure the potential difference across points 1 and 2 of the parallel combination of the batteries. Record your measured value.

Potential difference across B and C in parallel: ________ Volts

g. How does your measured value agree with your predictions?

23.3. POTENTIAL DIFFERENCES IN SERIES CIRCUITS

You can now explore the potential difference across different parts of a simple series circuit.

To do this you will need:

- 1 D-cell batteries, 1.5 V, alkaline (fresh)
- 1 D-cell holder
- 6 alligator clip leads, > 10 cm
- 2 #14 bulbs (with identical brightness)
- 2 #14 bulb holders
- 1 SPST switch
- 1 digital voltmeter
- 1 ammeter, 0.25 A

Recommended group size:	2	Interactive demo OK?:	N

Let's begin with the circuit with two bulbs in series with a battery that you looked at before in Unit 22. It is shown in Figure 23.5.

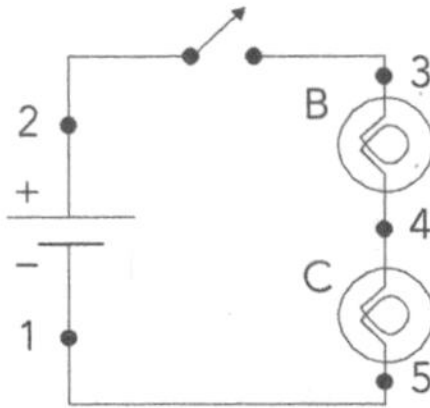

Fig. 23.5. A series circuit with one battery and two bulbs.

23.3.1. Activity: Voltages in Series Circuits

a. If bulbs B and C in Figure 23.5 are identical, how do you predict that the potential difference across bulb B (points 3 and 4) will compare to the potential difference (voltage) across the battery (points 1 and 2)? How about the potential difference across bulb C (points 4 and 5)? How will the potential difference across the series combination of bulbs B and C (points 3 and 5) compare to potential difference across the battery?

b. Test your predictions by measuring the voltages and recording their values.

Potential difference across the battery: ________________ Volts

Potential difference across bulb B: ____________________ Volts

Potential difference across bulb C: ____________________ Volts

Potential difference across bulbs B and C in series: ______ Volts

c. How do the three potential differences compare? Did your observations agree with your predictions?

d. Formulate a rule for how potential differences across individual bulbs in a series connection combine to give the total potential difference across the series combination of the bulbs. How is this related to the potential difference of the battery?

23.4. POTENTIAL DIFFERENCE IN PARALLEL CIRCUITS REVISITED

You can also explore the potential differences across different parts of a simple parallel circuit.

To do this you will need:

- 2 D-cell batteries, 1.5 V (fresh alkaline)
- 2 D-cell holders
- 6 alligator clip leads, > 10 cm
- 3 #14 bulbs (with identical brightness)
- 3 #14 bulb holders
- 1 SPST switch
- 1 digital voltmeter
- 1 ammeter, 0.25 A

Recommended group size:	2	Interactive demo OK?:	N

Let's begin with the circuit with two bulbs in parallel with a parallel battery setup similar to one you looked at before in Unit 22. This circuit is shown in Figure 23.6.

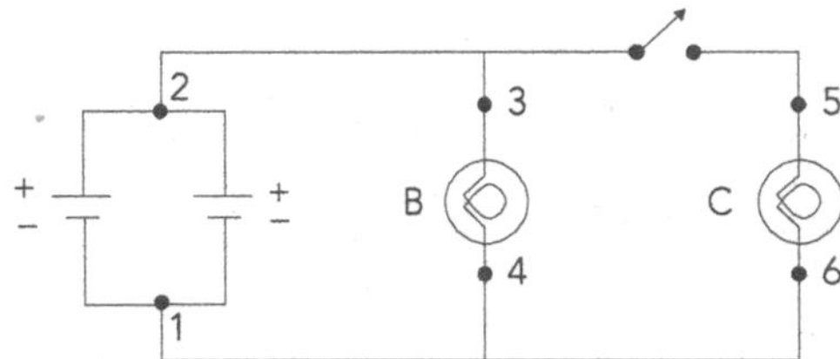

Fig. 23.6. Parallel circuit with two bulbs and a parallel battery setup.

23.4.1. Activity: Potential Differences in a Parallel Circuit

a. What do you predict will happen to the potential difference across the battery when you close the switch in Figure 23.6? Will it increase, decrease, or remain essentially the same?

b. How will the potential difference across bulb B compare to the voltage of the battery? How will the potential difference across bulb C compare to the voltage of the battery?

c. Measure the potential difference across the battery (points 1 and 2) and then across bulb B (points 3 and 4) with the switch open and closed. List your results to one decimal place (tenths of a volt) as we are not interested in differences to hundredths of a volt.

	Voltage across battery	Voltage across bulb B
Switch open		
Switch closed		

Note: Please round your results to one decimal place.

d. Did your measurements agree with your predictions? Did closing and opening the switch significantly affect the voltage across the battery? The voltage across bulb B?

e. Measure the potential difference across bulb B (points 3 and 4) and then across bulb C (points 5 and 6) with the switch open and closed.

	Voltage across bulb B	Voltage across bulb C
Switch open		
Switch closed		

Note: Please round your results to one decimal place.

f. Did your measurements agree with your predictions? Did closing and opening the switch significantly affect the voltage across bulb B?

g. Did closing and opening the switch significantly affect the voltage across bulb C? Under what circumstances is there a potential difference across a bulb?

h. Based on your observations, formulate a rule for the potential differences across the branches of a parallel circuit. How are these related to the voltage across the battery?

i. Based on your observations in this and the last two activities, is the potential difference across a fresh alkaline battery significantly affected by the circuit connected to it?

j. Is a battery a constant current source (delivering a fixed amount of current regardless of the circuit connected to it) or a constant voltage source (applying a fixed potential difference regardless of the circuit connected to it), or neither? Explain based on your observations in this and previous units.

23.5. APPLYING CURRENT AND POTENTIAL DIFFERENCE MODELS

The next two activities are designed to help you apply and consolidate what you have learned. For these you will need:

- 3 D-cell batteries, 1.5 V, alkaline (fresh)
- 3 D-cell holders
- 6 alligator clip leads, > 10 cm
- 3 #14 bulbs (with identical brightness)
- 3 #14 bulb holders
- 1 SPST switch

Recommended group size:	2	Interactive demo OK?:	N

Potential Differences Across Circuit Wire

So far you have been measuring voltages across circuit elements like a battery, which is a circuit element, and a light bulb, which has a resistance to the flow of electricity. Is there a potential difference across wiring that is used in a circuit?

Let's consider this question by making some more voltage drop measurements on the circuit depicted in Figure 23.6.

23.5.1. Activity: Potential Difference Across Wire

a. What do you predict will be the potential difference across two points of wire when no circuit element is present? For example, how about the voltage across the stretch of wire that connects each terminal of the battery directly to bulb B (points 2 and 3 or points 1 and 4)? How about the potential difference across points 1 and 6 when the switch is closed?

b. Check your prediction by measuring the potential differences between points 2 and 3 and points 1 and 4. Also measure the potential difference across points 1 and 6 with the switch closed.

Potential difference across points 2 and 3: ________ Volts

Potential difference across points 1 and 4: ________ Volts

Potential difference across points 1 and 6: ________ Volts

c. What do you conclude about the potential difference—that is, the potential difference between two points on a conducting wire when no circuit element lies in between. Does circuit wire have much resistance? Explain.

Return to a Complex Circuit

In Unit 22 you explored what happened to the brightness of the bulbs in the circuit shown below when the switch was closed—that is, when bulb D was added in parallel with bulb C.

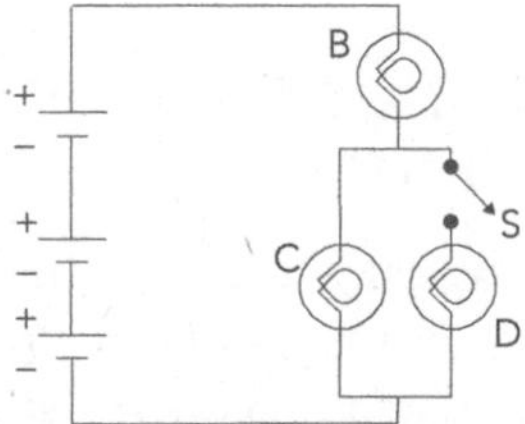

Fig. 23.7. Circuit with bulbs B and C in series when the switch is open and with bulb B in series with the parallel combination of bulbs C and D when the switch is closed.

You were previously asked to rank the brightness of bulbs B, C, and D after the switch was closed. The question now is *what happens to the brightness of bulb B when the switch is closed?* Does it increase, decrease, or remain the same?

23.5.2. Activity: Applying Current and Voltage Models

a. Based on the current and voltage models you have developed, *carefully* predict what will happen to the current through bulb B (and therefore its brightness) when bulb D is added in parallel to bulb C—will the brightness increase, decrease, or remain the same? Why?

b. Connect the circuit in Figure 23.7 and make observations. Describe what happens to the brightness of bulb B when the switch is closed.

c. Did your observations agree with your prediction? If not, use the current and voltage models to explain your observations.

OHM'S LAW, RESISTANCE, AND NETWORKS

23.6. OHM'S LAW: RELATING POTENTIAL DIFFERENCE, CURRENT, AND RESISTANCE

You have already seen on several occasions that there is only a potential difference across a bulb when there is a current flowing through a bulb. In the next activity we are going to use a carbon resistor instead of a light bulb. Resistors are designed to have the same resistance value no matter how much current is passing through. How does the potential difference across a resistor depend on the current through it? In order to explore this, you will need:

- 4 D-cell batteries, 1.5 V, alkaline
- 4 D-cell holders
- 1 multimeter
- 1 ammeter, 0.25 A
- 1 ammeter, 1 A
- 1 resistor, approximately 75Ω
- 1 SPST switch

Recommended group size:	2	Interactive demo OK?:	N

23.6.1. Activity: Experimental Relationship of i and ΔV

a. What do you predict will happen to the voltage drop, ΔV, *across* the resistor as the current, i, *through* it increases? Sketch a graph of your predicted relationship.

Fig. 23.8. Circuit diagram resistor symbol.

b. Set up a circuit to test your prediction by placing the resistor in series with one, two, three, and then four batteries. Set up the voltmeter and ammeter to measure the voltage across the resistor and the current *through* it. Carefully describe your procedures and results, and sketch your circuit diagram. **Note:** The circuit symbol for a resistor is slightly different from that for a bulb. It is shown in Figure 23.8.

c. Summarize your data for ΔV as i increases in the data table.

Number of batteries	i (Amps)	ΔV (Volts)
0	0.00	0.00
1		
2		
3		
4		

d. Create a computer-generated graph of ΔV vs. i and affix a printout of it in the space below. What is the significance of the slope of the graph?

e. How does the shape of your graph compare with what you predicted in part a?

Defining Resistance

We define a quantity known as resistance for a circuit element as

$$R \equiv \frac{\Delta V}{i}$$

where ΔV is the potential difference across a circuit element and i is the current through it.

If potential difference is measured in volts and current is measured in amperes, then the unit of resistance is the ohm, which is usually represented by the Greek letter Ω, "omega."

23.6.2. Activity: Statement of Ohm's Law

a. State the mathematical relationship found in Activity 23.6.1 between potential difference and current for a resistor in terms of ΔV, i, and R.

b. Based on your graph, what can you say about the value of R for a resistor—is it constant or does it change as the current through the resistor changes? Explain.

c. From the slope of your graph, what is the experimentally determined value of the resistance of your resistor in ohms? How does your slope agree with the rated value of the resistor? If you use a resistor rated at 75Ω, the slope should be close to 75Ω but is unlikely to be exactly 75Ω.

Fig. 23.9.

d. Complete the famous pre-exam rhyme used by countless introductory physics students throughout the English-speaking world:

Twinkle, twinkle little star, ΔV equals ________ times ________

(1) If a circuit element (such as the carbon resistor you have been using) has the same resistance over a wide range of conditions, it is called *ohmic*. This is because it obeys Ohm's Law with ΔV and i being proportional to each other. (2) A circuit element like a light bulb that changes resistance with the amount of current in it is called *non-ohmic*.

Note: The resistance for a non-ohmic resistor is still determined by $R = \Delta V/i$ but R is not constant. Instead, it depends on ΔV.

23.7. USING A MULTIMETER

A digital multimeter is a device that can be used to measure either current, voltage, or resistance depending on how it is set up. The following activity will introduce you to the digital multimeter and give you some practice in using it. You will need:

- 1 multimeter
- 1 D-cell battery, 1.5 V, alkaline
- 1 D-cell holder
- 1 SPST switch
- 4 alligator clip leads, > 10 cm
- 1 resistor, 10 Ω

Recommended group size:	2	Interactive demo OK?:	N

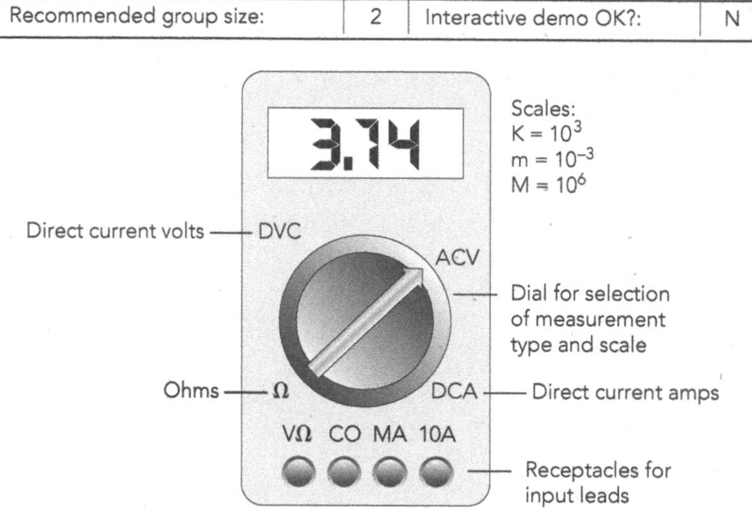

Fig. 23.10. Diagram of a digital multimeter that can be used to measure resistances, currents, and voltages.

By putting the input leads (red for positive, black for negative) into the proper receptacles and setting the dial correctly, you can measure resistances (Ω) as well as direct currents (DCA) and direct current voltages (DCV).

Figure 23.11 shows a simple circuit that you can use to practice taking readings with the multimeter.

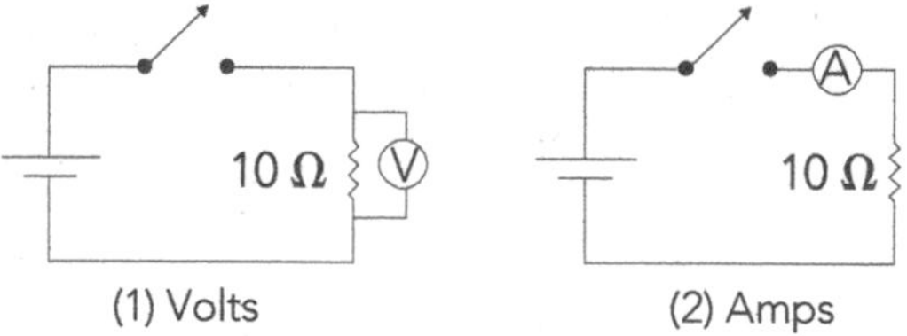

Fig. 23.11. Simple circuit for using a multimeter to measure (1) volts or (2) amps.

23.7.1. Activity: Using a Multimeter

a. Determine what settings you need to use to measure the actual resistance of the "10Ω" resistor. Record your measured value below. **Note:** Resistances must be measured when no currents are present. It is best to take resistors out of the circuit before measuring them.

b. Determine what settings you need to measure the potential difference across a resistor in series with a battery. Connect the circuit in Figure 23.11. Close the switch and measure the voltage across the resistor. Record the voltage below.

c. Reconnect the circuit, this time using the multimeter as an ammeter. Close the switch again and measure the current *through* the resistor. Record the current below.

23.8. RESISTANCE AND ITS MEASUREMENT

In the series of observations you have been making with batteries and bulbs it is clear that electrical energy is being transferred to light and heat energy inside a bulb, so that even though all the current returns to the battery after flowing through the bulb, the charges have lost potential energy. We say that when electrical potential energy is lost in part of a circuit, such as it is in the bulb, it is because that part of the circuit offers *resistance* to the flow of electric current.

A battery causes charge to flow in a circuit. The electrical resistance to the flow of charge can be compared to the mechanical resistance offered by the pegs and the barrier in a mechanical model depicted by a ramp with bowling balls traveling down it.

A light bulb is one kind of electrical resistance. Another common kind of electrical resistance is provided by a carbon resistor manufactured to provide a constant resistance in electrical circuits.

Carbon resistors are the most standard sources of resistance used in electrical circuits for several reasons. A light bulb has a resistance that increases with temperature and current and thus doesn't make a good circuit element when quantitative attributes are important. The resistance of carbon resistors doesn't vary with the amount of current passing through them. Carbon resistors are inexpensive to manufacture and can be produced with low or high resistances.

A typical carbon resistor contains a form of carbon, known as graphite, suspended in a hard glue binder. It usually is surrounded by a plastic case with a color code painted on it. It is instructive to look at samples of carbon resistors that have been cut down the middle as shown in Figure 23.12.

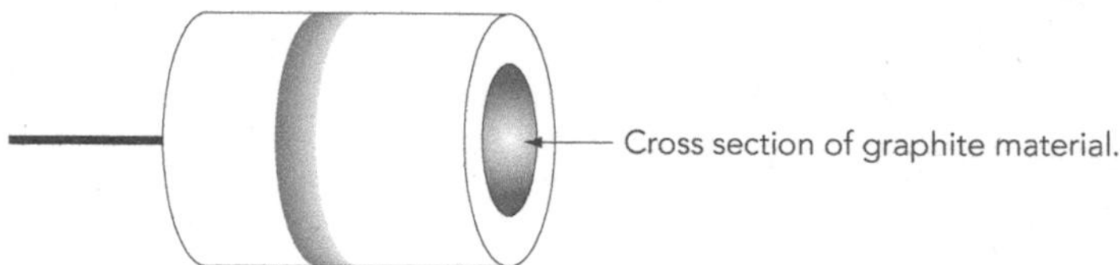

Fig. 23.12. A cutaway view of a carbon resistor showing the cross-sectional area of the graphite material.

As you found in the previous activity on Ohm's law, a simple equation can be used to define electrical resistance in terms of potential difference, ΔV, across it and the current, i, through it. It is

$$R \equiv \frac{\Delta V}{i} \qquad \text{(definition of } R\text{)}$$

Note: For carbon resistors this R will be a constant for a range of different currents. This is not true for filament light bulbs. Their R-value increases as the current is increased because the bulb is heating up. Light bulbs are poor ohmic resistors.

If the potential difference in a circuit is measured in volts and the current in amperes, the resistance is assigned the unit of *ohms* represented by the Greek letter Ω (omega).

A carbon resistor is usually marked with colored bands to signify its value in ohms.

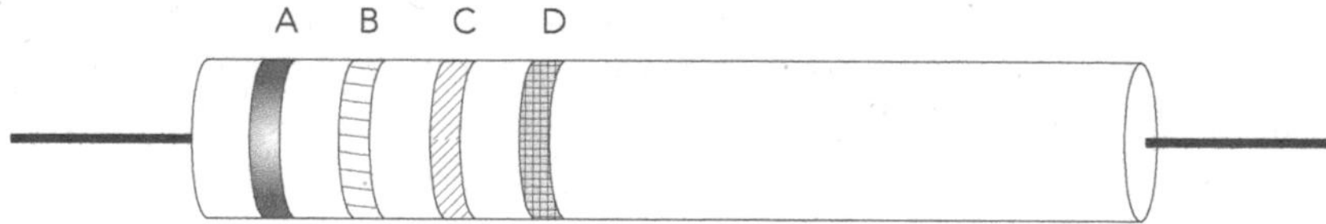

Fig. 23.13. A carbon resistor with color bands.

Each color has a number associated with it. These associations are shown in the following table.

A table representing the resistor code	
Black = 0	Blue = 6
Brown = 1	Violet = 7
Red = 2	Gray = 8
Orange = 3	White = 9
Yellow = 4	Silver = ±10%
Green = 5	Gold = ±5%

The value in ohms $= AB \times 10^{C} \pm D$. (AB means the A digit placed beside the B digit, not A times B.) The colors on bands A, B, and C represent the digits shown in the table above. The D band represents the "tolerance" of the resistor. No band denotes ±20%, a silver band denotes ±10%, and a gold band denotes ±5%.

For example, a resistor with bands of Yellow-Gray-Red-Silver has a value: $AB \times 10^c \pm D = 48 \times 10^2 \pm 10\%$ ohms or $4800 \pm 480\ \Omega$, since $A = 4$, $B = 8$, $c = 2$, $D =$ silver ($\pm 10\%$).

Remembering the Code: There are a number of ditties that have been devised to help people remember the resistor code. Some of them are too "colorful" for official publication and others are too boring. A good compromise is found in the ditty *"Bad Booze Rots Our Young Guts But Vodka Goes Well"* in which the BBROYGBVGW sequence of first letters matches that for Black, Brown, Red, etc.

Suppose you have finally graduated and taken a job as a quality control inspector for a company that makes carbon resistors. Your task is to determine the rated resistance in ohms of a batch of five carbon resistors and then check your decoding skills by measuring the resistance with a digital multimeter.

For this activity you'll need:

- 5 assorted color-coded carbon resistors
- 1 multimeter

Recommended group size:	2	Interactive demo OK?:	N

To use your multimeter to measure resistance, flip the indicator switch to the section marked resistance (or ohms or Ω). Then connect the black multimeter probe to the common input and the red probe to the input designated for measuring resistance.

Finally, decode each resistor using the color bands on it, and then determine the discrepancy in percentage between the coded value of resistance and the value you determine experimentally. In general, percentage discrepancy is given by the expression

$$\text{Percent discrepancy} = \left| \frac{\text{Accepted value} - \text{Experimental value}}{\text{Accepted value}} \right| \times 100$$

23.8.1. Activity: Decoding and Measuring Resistors

a. Decode the five resistors, determine the resistance of each, and fill in the following table.

Color sequence	Coded R (Ω)	Measured R (Ω)	Calculated percent discrepancy	Percent tolerance	Within rated tolerance
1.					
2.					
3.					
4.					
5.					

b. Would you buy resistors from the manufacturer who made the batch you tested? Why or why not?

23.9. CARBON RESISTORS IN PARALLEL AND SERIES

Since the resistance of carbon resistors doesn't change as current is increased, they are often used in electrical circuits. Not all resistors are made of carbon. Since the resistance of a wire with a uniform cross-sectional area is directly proportional to length, it is possible to control the R-value of a wire fairly precisely. Thus, precision resistors with good temperature stability are often made of windings of fine wire.

Several identical resistors can be wired in series to increase their effective length and in parallel to increase their effective cross-sectional area as shown in Figure 23.14.

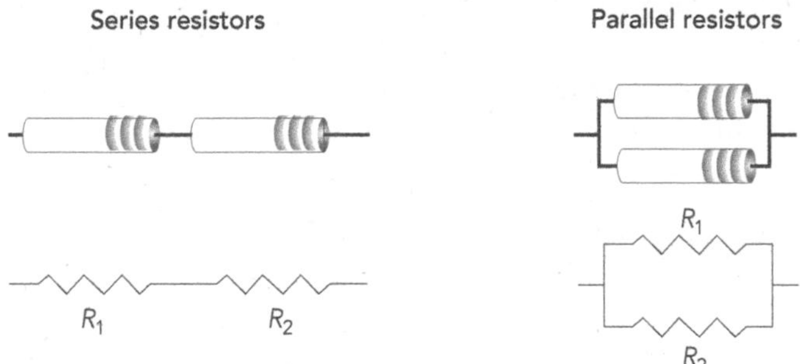

Fig. 23.14. Carbon resistors wired in series and in parallel.

In order to test predictions and explore the nature of equivalent resistances of different combinations of resistors you will need the following:

- 3 carbon resistors, 100 Ω
- 3 carbon resistors, 220 Ω
- 1 multimeter
- 4 alligator clip leads, approx. 5 cm

Recommended group size:	2	Interactive demo OK?:	N

Use the items listed to devise a way to measure the equivalent resistance when three or more resistors are wired in series. Explain what you did and summarize the results of your measurements.

23.9.1. Activity: Resistances for Series Wiring

a. If you have three different carbon resistors, what do you think the equivalent resistance to the flow of electrical current will be if the resistors are wired in *series*? Explain the reasons for your prediction based on your previous observations with batteries and bulbs.

b. Compare the calculated and measured values of equivalent resistance of the series network as follows:

Write down the *measured* values of each of the three resistors:

$R_1 =$ ________ Ω

$R_2 =$ ________ Ω

$R_3 =$ ________ Ω

Describe the method you are using to predict the equivalent resistance and calculate the predicted R-value:

Predicted $R_{eq} =$ ________ Ω

c. Draw a diagram for the resistance network for the three *different* resistors *wired in series.* Mark the measured values of the three resistances on your diagram.

d. Measure the actual resistance of the series resistor network and record the value:

Measured R_{eq} = ________ Ω

e. How does this value compare with the one you predicted on the basis of your calculation?

f. On the basis of your experimental results, devise a general mathematical equation that describes the equivalent resistance when n resistors are wired in series. Use the notation R_{eq} to represent the equivalent resistance and $R_1, R_2, R_3, \ldots R_n$ to represent the values of the individual resistors.

Devise a way to measure the equivalent resistance when two or more resistors are wired in *parallel*. Explain what you did and summarize the results of your measurements. Draw a symbolic diagram for each of the wiring configurations you use.

23.9.2. Activity: Resistances for Parallel Wiring

a. If you have two *identical* carbon resistors, what do you think the resistance to the flow of electrical current will be if the resistors are wired in *parallel*? Explain the reasons for your prediction.

b. Pick out two carbon resistors with an identical color code and draw a diagram for these two resistors *wired in parallel*. Label the diagram with the measured values R_1(measured) and R_2(measured). Predict the equivalent resistance of the parallel circuit and record your prediction below. Measure the value of the equivalent resistance of the network. Explain your reasoning and show your calculations in the space below.

Predicted value: R_{eq} = ________ Ω

Measured value: R_{eq} = ________ Ω

c. Pick out three *different* resistors and draw a diagram for these three resistors *wired in parallel.* Label the diagram with the measured values R_1(measured), R_2(measured), and R_3(measured). Measure the value of the equivalent resistance of the network and record it below.

Measured value of the equivalent resistance of the network:

R_{eq} = ________ Ω

d. Use the notation R_{eq} to represent the equivalent resistance and R_1, R_2, R_3, . . . , etc. to represent the values of the individual resistors. Show that, within the limits of experimental uncertainty, the results of the measurements you made with parallel resistors are the same as those calculated using the equation:

$$\frac{1}{R_{eq}} = \frac{1}{R_1} + \frac{1}{R_2} + \frac{1}{R_3} + \ldots$$

1. For the two identical resistors wired in parallel:

Calculated value: R_{eq} = ________ Ω

Measured value: R_{eq} = ________ Ω

2. For the three resistors wired in parallel:

Calculated value: $R_{eq} =$ ________ Ω

Measured value: $R_{eq} =$ ________ Ω

e. Show mathematically that if

$$\frac{1}{R_{eq}} = \frac{1}{R_1} + \frac{1}{R_2} \quad \text{then} \quad R_{eq} = \frac{R_1 R_2}{(R_1 + R_2)}$$

23.10. EQUIVALENT RESISTANCES FOR NETWORKS

Now that you know the basic equations to calculate equivalent resistance for series and parallel resistances, you can tackle the question of how to find the equivalent resistances for complex networks of resistors. The trick is to be able to calculate the equivalent resistance of each segment of the complex network and use that in calculations of the next segment. For example, in the network shown below there are two resistance values R_1 and R_2. A series of simplifications is shown in Figure 23.15.

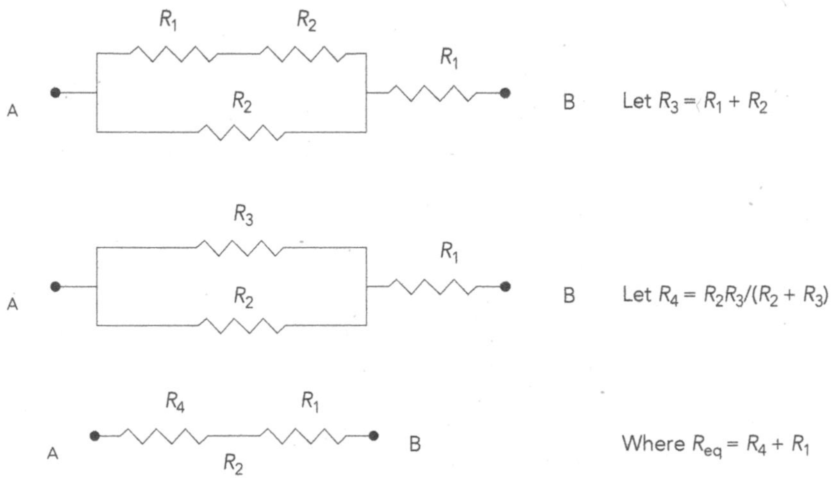

Fig. 23.15. A sample resistor network.

In order to complete the equivalent resistance activities you will need the following apparatus:

- 3 carbon resistors, 100 Ω
- 3 carbon resistors, 220 Ω
- 1 multimeter

Recommended group size:	2	Interactive demo OK?:	N

23.10.1. Activity: The Equivalent Resistance for a Network

a. Consider the sets of identical resistors you just used to explore parallel and series resistances. Use the color-coded value for your lowest identical resistor for R_1 and the color-coded value for your highest identical resistor for R_2 to calculate the equivalent resistance between points A and B for the network shown below. *You must show your calculations on a step-by-step basis.*

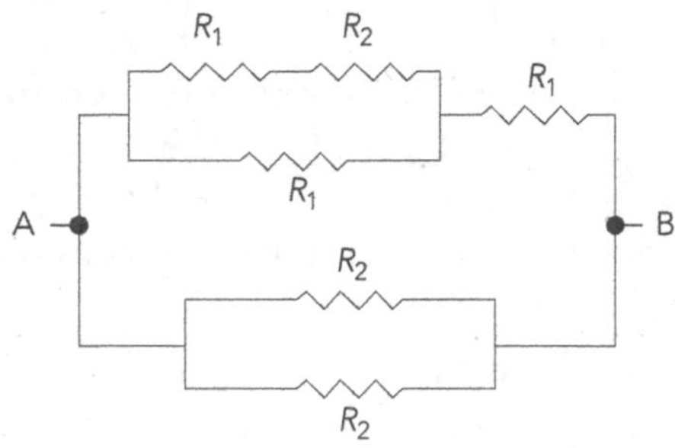

Fig. 23.16.

b. Set up the network of resistors and check your calculation by measuring the equivalent resistance directly.

Calculated value: R_{eq} = ________ Ω

Measured value: R_{eq} = ________ Ω

CONFIRMING KIRCHHOFF'S LAWS

23.11. KIRCHHOFF'S LAWS

Suppose we wish to calculate the currents in various branches of a circuit that has many components wired together in a complex array. In such circuits, simplification using series and parallel combinations is often impossible. Instead we can state and apply a formal set of rules known as Kirchhoff's laws to use in the analysis of current flow in circuits. These rules are:

Kirchhoff's Laws

1. *Junction (or node) Rule (based on charge conservation)*: The sum of all the currents entering any node or branch point of a circuit (that is, where two or more wires merge) must equal the sum of all currents leaving the node.
2. *Loop Rule (based on energy conservation)*: Around any closed loop in a circuit, the sum of all emf, voltage gains provided by batteries or other power sources, ($\mathscr{E}$ = emf) and all the potential drops across resistors and other circuit elements must equal zero.

Steps for Applying Rules

1. Assign a current symbol to each branch of the circuit and label the current in each branch i_1, i_2, i_3, etc.; then *arbitrarily* assign a direction to each current. (The direction chosen for the circuit for each branch doesn't matter. If you chose the "wrong" direction, the value of the current will simply turn out to be negative.) Remember that the current flowing out of a battery is always the same as the current flowing into a battery.
2. Apply the loop rule to each of the loops by: (a) letting the potential drop across each resistor be the negative of the product of the resistance and the net current through that resistor (reverse the sign to "plus" if you are traversing a resistor in a direction opposite that of the current); (b) assigning a positive potential difference when the loop traverses from the − to the + terminal of a battery. (If you are going through a battery in the opposite direction, assign a *negative* potential difference to the trip across the battery terminals.)
3. Find each of the junctions and apply the junction rule to it. You can place currents leaving the junction on one side of the equation and currents coming into the junction on the other side of the equation.

In order to illustrate the application of the rules, let's consider the circuit in Figure 23.17.

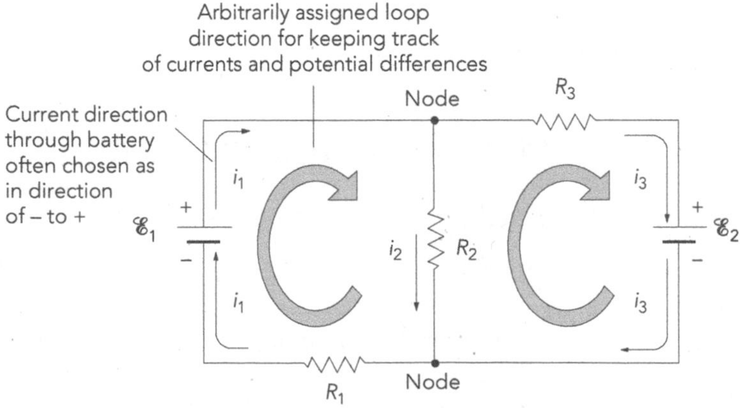

Fig. 23.17. A complex circuit in which loops 1 and 2 share the resistor R_2. The currents i_1 and i_3 flow through R_2 in opposite directions and the net current through R_2 is denoted by i_2.

In Figure 23.17, the directions for the currents through the branches and for i_2 are assigned arbitrarily. *If we assume that the internal resistances of the batteries are negligible,* then by applying the loop and junction rules we find that

Loop 1 Eq.:	$\mathscr{E}_1 - i_2 R_2 - i_1 R_1 = 0$	(23.1)
Loop 2 Eq.:	$-\mathscr{E}_2 + i_2 R_2 - i_3 R_3 = 0$	(23.2)
Node 1 Eq.:	$i_1 = i_2 + i_3$	(23.3)

(current into junction = current out of junction)

It is not obvious that the loops and their directions can be chosen arbitrarily. Let's explore this assertion theoretically for a simple situation and then more concretely with some specific calculations. In order to do the following activity you'll need a couple of resistors and a multimeter as follows:

- 1 carbon resistor, 39 Ω
- 1 carbon resistor, 75 Ω
- 1 multimeter
- 1 battery, 4.5 V
- 1 D-cell battery, 1.5 V, alkaline

Recommended group size:	4	Interactive demo OK?:	N

23.11.1. Activity: Applying the Loop Rule Several Times

a. Use the loop and node rule along with the new arbitrary direction for I_2 to rewrite the three equations relating values of battery emfs, resistance, and current in the circuit shown in Figure 23.18.

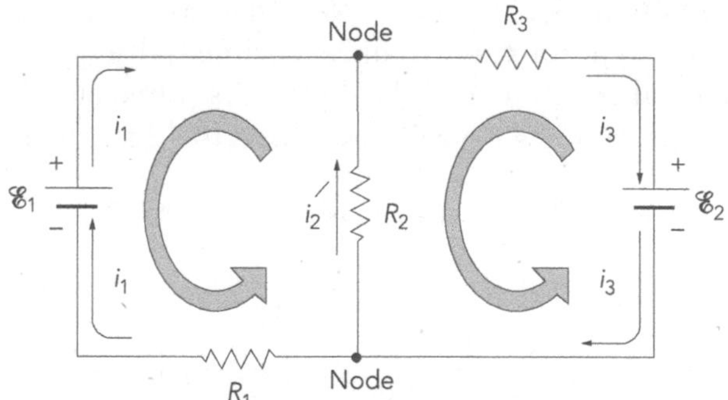

Fig. 23.18. A circuit identical to that in Fig. 23.17 with initial assumptions that: (1) current flows in the opposite direction through R_2; (2) arbitrarily assigned loops are taken in opposite directions.

b. Show that if $i_2' = -i_2$ then the three equations you just constructed can be rearranged algebraically so they are *exactly the same* as Equations 23.1, 23.2, and 23.3.

c. Suppose the values of each component for the circuit shown in Figure 23.18 are rated as

$$\mathcal{E}_1 = 4.5 \text{ V}$$
$$\mathcal{E}_2 = 1.5 \text{ V}$$

Rated fixed resistances: $R_1 = 75\ \Omega$

$R_3 = 39\ \Omega$

Variable resistance: $R_2 = 100\ \Omega$

1. Since you are going to test your theoretical results for Kirchhoff's law calculations for this circuit experimentally, you should measure the actual values of the two fixed resistors (rated at 75 Ω and 39 Ω) and the two battery voltages with a multimeter. List the results below.

Measured value of the battery emf rated at 4.5 V $\mathcal{E}_1 =$ ________V

Measured value of the battery emf rated at 1.5 V $\mathcal{E}_2 =$ ________V

Measured value of the resistor rated at 75 Ω $R_1 =$ ________Ω

Measured value of the resistor rated at 39 Ω $R_3 =$ ________Ω

2. Carefully rewrite Equations 23.1, 23.2 and 23.3 with the appropriate *measured* (not rated) values for emf and resistances substituted into them. Use 100 Ω for the value of R_2 in your calculation. In the experiment, you will be setting a 200 Ω variable resistor to a value of 100 Ω.

d. Solve these three equations for the three unknowns i_1, i_2, and i_3 in amps using one of the following methods: (1) substitution, (2) determinants, or (3) equation-solving computer or programmable calculator software (see Appendix I for details on computer or calculator solutions). Either show your calculations or name and describe the computer or calculator tool you used in the space below. If you use a computer tool, include a printout of the command and the solution.

e. Show by substitution that your solutions actually satisfy your first equation, 23.1.

23.12. VERIFYING KIRCHHOFF'S LAWS EXPERIMENTALLY

Since circuit elements have become smaller in the past 20 years or so, it is common to design and wire simple circuits on a device called a protoboard. A protoboard has hundreds of little plastic holes in it that can have small-diameter wire poked into them. In the protoboard model shown in Figure 23.19, these holes are electrically connected in vertical columns of 5 near the middle. The top of the protoboard has two horizontal rows of 40 connected holes. There is a similar arrangement at the bottom.

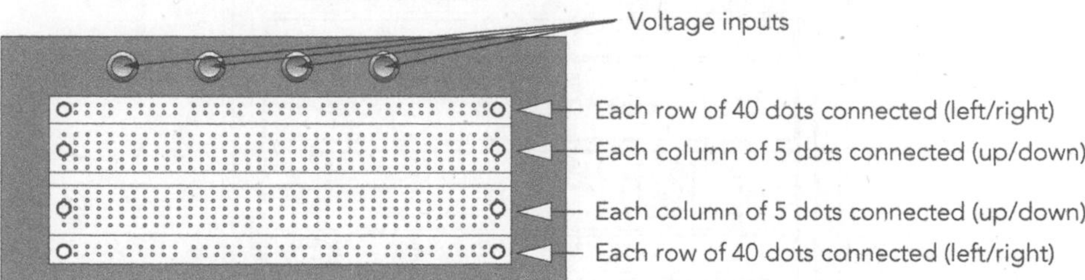

Fig. 23.19. A protoboard.

Usually, one connects the voltage inputs to the long rows of connected dots toward the outside of the circuit; these rows can then serve as power supplies. As part of the next project with the protoboard you will be using some simple circuit elements to design a tricky circuit with more than one battery and several branches in it. To design this circuit you will be using the following items:

- 1 pot, 200 Ω (variable resistance set at 100 Ω)
- 1 carbon resistor, 39 Ω
- 1 carbon resistor, 75 Ω
- 4 D-cell batteries (fresh)
- 1 protoboard
- 1 multimeter
- 1 small screwdriver
- Assortment of small lengths of #22 wire (for use w/ the protoboard)

Recommended group size:	2	Interactive demo OK?:	N

The word "pot" stands for potentiometer. It is a variable resistor. There is a 200-Ω pot already installed on your protoboard. The pot has three leads. The two outside leads are across the 200-Ω resistor while the center lead taps off part of the 200 Ω. The resistance between an outside lead and the center tap can be adjusted from 0 to 200 Ω with a screwdriver; the resistance between the two outside leads is always 200 Ω. The circuit symbol for the pot is shown below.

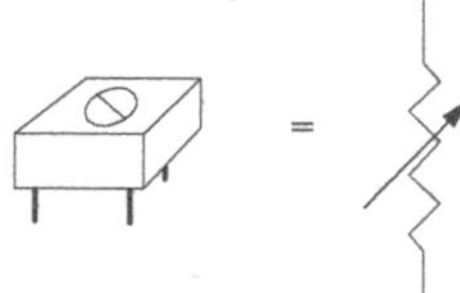

Fig. 23.20. A protoboard pot.

To wire up the circuit shown in Figure 23.21 on the protoboard, you will need to examine the details of how the protoboard is arranged, as shown in Figure 23.19. Although there are many legitimate ways to connect the leads, a possible configuration is shown in the figure below.

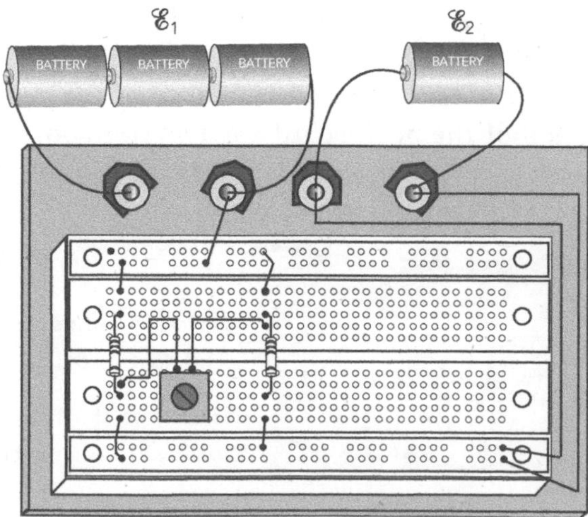

Fig. 23.21. A possible wiring arrangement for the Kirchhoff's law circuit to be tested.

23.12.1. Activity: Testing the Loop Rule with a Real Circuit

a. Use the ohmmeter feature of the digital multimeter to measure the total resistance across a pot that is labeled 200 Ω. Then measure the resistance between the center tap on the pot and one of the other taps. What happens to the ohmmeter reading as you use a paper clip or small screwdriver to change the setting on the pot?

b. Set the pot so that there is 100 Ω between the center tap and one of the other taps. Was it difficult?

c. Wire up the circuit pictured in Figure 23.21; use a protoboard and the pot (set at 100 Ω) as R_2. Measure the current in each branch of the circuit and compare the measured and calculated values of the current by computing the percent discrepancy in each case. **Note:** Don't use an ammeter! The most accurate way to measure current with a digital multimeter is to measure the potential difference across each of the resistors and use Ohm's law to calculate i from ΔV and R.

	Measured R (Ω)	Measured ΔV (volts)	Measured $i = \Delta V/R$ (amps)	Theoretical i (amps)	Percent discrepancy
1.					
2.					
3.					

d. What do you predict will happen to each of the currents as the resistance on the pot is decreased? That is, will the currents i_1, i_2, and i_3 increase or decrease? Explain your predictions.

e. What actually happens to each of the currents as you decrease R_2? How good were your predictions?

Name ______________________ Section ____________ Date ____________

UNIT 24: CAPACITORS AND RC CIRCUITS

A capacitor consists of two conductors separated from each other by insulating material. They come in all sizes and shapes and can be found in the vast majority of electrical circuits. The photograph shows a large old-fashioned high-voltage capacitor with a new type of low-voltage capacitor held in front of it. Incredibly the tiny new capacitor has 1000 times the capacity of the old one. The name capacitor suggests a capacity to hold something, but what? How does it behave in a simple direct current circuit with a resistor? How are capacitors used in circuits in which currents change with time? When you complete this unit, you should be able to answer these questions.

UNIT 24: CAPACITORS AND RC CIRCUITS

The most universal and significant concept to come out of the work on the telegraph was that of capacitance. H. I. Sharlin

I get a real charge out of capacitors. P. W. Laws

OBJECTIVES

1. To define capacitance and learn how to measure it with a digital multimeter.

2. To discover how the capacitance of parallel plates is related to the area of the plates and the separation between them.

3. To determine how capacitance changes when capacitors are wired in parallel and when they are wired in series by using physical reasoning, mathematical reasoning, and direct measurements.

4. To discover how the charge on a capacitor and the electric current change with time when a charged capacitor is placed in a circuit with a resistor.

24.1. OVERVIEW

Any two conductors separated by an insulator can be electrically charged so that one conductor has a net excess positive charge and the other conductor has an equal amount of excess negative charge; such an arrangement is called a capacitor. A capacitor can be made up of two strange-shaped blobs of metal or it can have any number of regular symmetric shapes such as that of one hollow sphere inside another, or one hollow rod inside another.

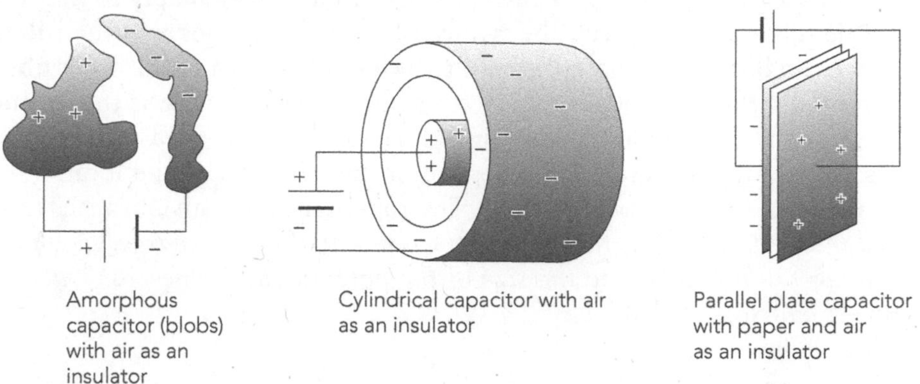

Fig. 24.1. Some different capacitor geometries.

The type of capacitor that is of the most practical interest is the parallel plate capacitor. Thus, we will focus exclusively on the study of the properties of parallel plate capacitors. There are a couple of reasons why you will be studying parallel plate capacitors. First, the parallel plate capacitor is the easiest to use when making mathematical calculations or using physical reasoning. Second, it is relatively easy to construct. Third, parallel plate capacitors are used widely in electronic circuits to do such diverse things as defining the flashing rate of a neon tube, determining what radio station will be tuned in, and storing electrical energy to run an electronic flash unit. Materials other than conductors separated by an insulator can be used to make a system that behaves like a simple capacitor. Although many of the most interesting properties of capacitors come in the operation of alternating current circuits, we will limit our present study to the properties of the parallel plate capacitor and its behavior in direct current circuits like those you have been constructing in the last couple of units. The circuit symbol for a capacitor is a pair of lines as shown in Figure 24.2.

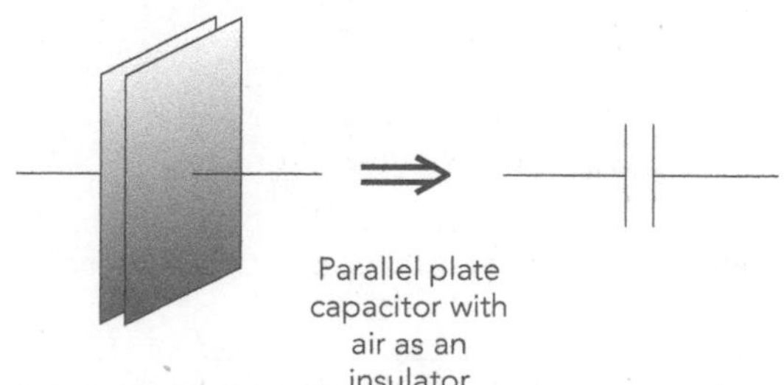

Fig. 24.2. The circuit diagram symbol for a capacitor.

CAPACITANCE, AREA, AND DISTANCE

24.2. THE PARALLEL PLATE CAPACITOR

The typical method for transferring equal and opposite excess charges to a capacitor is to use a voltage source such as a battery or power supply to impress a potential difference between the two conductors. Electrons will then flow off of one conductor (leaving excess positive charges) and on to the other until the potential difference between the two conductors is the same as that of the voltage source. In general, the amount of excess charge needed to reach the impressed potential difference will depend on the size, shape, and location of the conductors relative to each other. The capacitance of a given capacitor is *defined mathematically as the ratio of the amount of excess charge, $|q|$, on either one of the conductors to the size of the potential difference, $|\Delta V|$, across the two conductors* so that

$$C \equiv \frac{|q|}{|\Delta V|} \tag{24.1}$$

Thus, capacitance is *defined* as a measure of the amount of net or excess charge on either one of the conductors per unit potential difference (or Coulombs per volt in SI units).

You can draw on some of your experiences with electrostatics to think about what might happen to a parallel plate capacitor when it is hooked to a battery as shown in Figure 24.3. This thinking can give you an intuitive feeling for the meaning of capacitance. For a fixed potential difference from a battery, the net excess charge found on either plate is proportional to the capacitance of the pair of conductors.

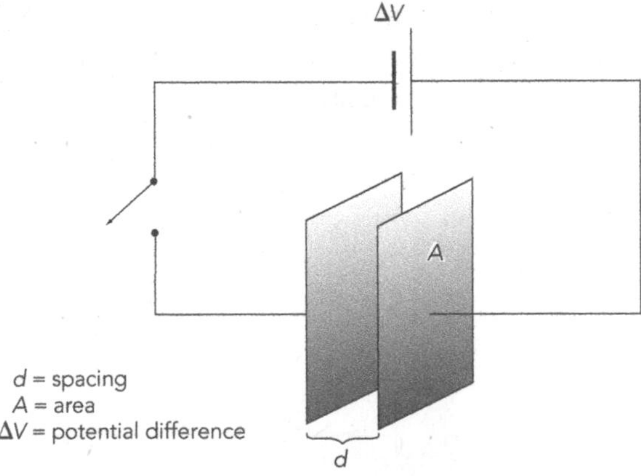

Fig. 24.3. A parallel plate capacitor with a potential difference ΔV across it.

24.2.1. Activity: Predicting Dependence on Area and Separation

a. Consider two identical metal plates of area A, separated by a nonconducting material that has a thickness d. They are connected in a circuit with a battery and a switch, as shown above. When the switch is open, there is no net excess charge on either plate. The switch is then closed. What will happen to the amount of net excess charge on the metal plate that is attached to the negative terminal of the battery? What will happen to the amount of net excess charge on the plate that is connected to the positive terminal of the battery? Explain.

b. Can excess charges on one plate of a charged parallel plate capacitor interact with excess charges on the other plate? If so, how? **Note:** To say that two charges interact is to say that they exert forces on each other from a distance.

c. Is there any limit to the amount of net excess charge that can be put on a plate? Explain.

d. Use qualitative reasoning to anticipate how the amount of net excess charge a pair of parallel plate conductors can hold will change as the area of the plates increases. Explain your reasoning.

e. Do you think that the amount of net excess charge a given battery can store on the plates will increase or decrease as the spacing, d, between the plates of the capacitor increases? Explain.

24.3. CAPACITANCE MEASUREMENTS FOR PARALLEL PLATES

The unit of capacitance is the farad, F, named after Michael Faraday. One farad is equal to one coulomb/volt. As you will demonstrate shortly, one farad is a very large capacitance. Thus, actual capacitances are often expressed in smaller units with alternate notation as shown below:

Units of capacitance
microfarad: $10^{-6} = 1\ \mu F$
picofarad: $10^{-12}\ F = 1\ pF = 1\ \mu\mu F$
nanofarad: $10^{-9} = 1\ nF = 1000\ \mu\mu F$

Note: Sometimes the symbol m is used instead of μ or *U* on capacitors to represent 10^{-6}, despite the fact that in other situations m always represents 10^{-3}!

Typically, there are several types of capacitors used in electronic circuits, including disk capacitors, foil capacitors, electrolytic capacitors, and so on. You might want to examine some typical capacitors. To do this you'll need:

- 4 capacitors (assorted collection)

To complete the next few activities you will need to construct a parallel plate capacitor and use a multimeter to measure capacitance. Thus, you'll need the following items:

- 2 pieces aluminum foil, 12 cm × 12 cm
- 1 textbook
- 1 multimeter (with capacitance measuring capability)
- 2 insulated wires (stripped at the ends, approximately 12" long)
- 1 ruler
- 1 Vernier caliper

Recommended group size:	4	Interactive demo OK?:	N

You can make a parallel plate capacitor out of two rectangular sheets of aluminum foil separated by pieces of paper. A textbook works well as the separator for the foil since you can slip the two foil sheets between any number of sheets of paper and weight the book down with something heavy and non-conducting like another massive textbook. You can then use your digital multimeter in its capacitance mode for the measurements. **Note:** Insert short wires into the capacitance slots of your multimeter as "probes"

When you measure the capacitance of your "parallel plates," be sure the aluminum foil pieces are arranged carefully so they don't touch each other and "short out."

24.3.1. Activity: Measuring How Capacitance Depends on Area or on Separation

a. Devise a way to measure how the capacitance depends on either the foil area or on the separation between foil sheets. If you hold the area constant and vary separation, record the dimensions of the foil so you can calculate the area. Alternatively, if you hold the distance constant, record its value. Take at least five data points in either case. Describe your methods and then create a data table with proper units and display a graph of the results.

b. Is your graph a straight line? If not, you should make a guess at the functional relationship it represents and create a model that matches the data. Affix your overlay graph showing both the data and your model in the space the follows. Be sure to label your graph axes properly.

c. What is the function that best describes the relationship between spacing and capacitance or between area and capacitance? How do the results compare with your prediction based on physical reasoning?

d. Use the ohmmeter on a piece of paper in your book. What is its resistance? Can current flow through the pages of your book?

24.4. DERIVING A MATHEMATICAL EXPRESSION FOR CAPACITANCE

We can use Gauss' law and the relationship between potential difference, ΔV, and electric field to derive an expression for the capacitance, C, of a parallel plate capacitor in terms of the area, A, and separation, d, of the aluminum plates. The diagram in Figure 24.4 is useful in this regard.

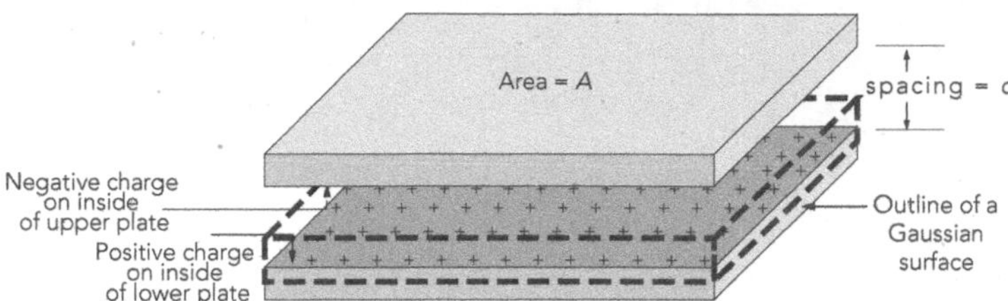

Fig. 24.4. A charged capacitor. The dotted lines outline a rectangular Gaussian surface enclosing positive charge. Area of the plate inside the Gaussian surface is ⊥ to the page.

24.4.1. Activity: Derivation of Capacitance vs. *A* and *d*

a. Write down the integral form of Gauss' law. However, use a new form of Gauss' law where the electric constant ε_0 (used when the material between the metal plates is just air) is replaced with a new constant $\varepsilon = \kappa\varepsilon_0$. In this case, κ (kappa) represents the dielectric constant of the pages in the textbook you are using as a spacer.

b. Examine the Gaussian surface shown in the diagram above. What is the value of the electric field *inside* the *bottom* plate? *What is the value of the electric flux through the bottom surface of the Gaussian surface?* **Hint:** There is never an electric field inside a conductor unless an electric current is present!

c. Using the results above and the notation $E = |\vec{E}|$ to represent the magnitude of the positive uniform electric field *between* the two plates, *find the net flux through the six surfaces of the Gaussian surface and set the total flux equal to the charge enclosed.* Next, show that $|q| = \varepsilon EA = \kappa\varepsilon_0 EA$ where the net positive charge on the bottom plate is denoted by q. **Note:** We are assuming that since the positive charges on the bottom plate are attracted to the negative charges on the top plate, the excess charges are on the inside surface of each plate.

d. Remember that in a uniform electric field the size of the potential difference (or voltage drop) across a distance d is given by $|\Delta V| = |E|d$. Use this fact and the definition of capacitance to show that

$$C = \frac{\kappa\varepsilon_0 A}{d}$$

where A is the area of the plates and d is their separation.

e. Use one of your actual areas and spacings from the measurements you made in Activity 24.3.1a to calculate a value of C. Assume that the dielectric constant, κ, for paper is about 3.5. How does the calculated value of C compare with the directly measured value? **Note:** κ (kappa) is a quantity called the dielectric constant and is a property of the insulating material that separates the two plates. For air, $\kappa = 1$. It is usually greater than 1 for other materials.

f. Now for an unusual question. If you have two square foil sheets, separated by paper with a dielectric constant of 3.5 that is 1 mm thick, how long (in miles) would each side of the sheets have to be in order to have $C = 1$ F? Show your calculations **Warning:** Miles are *not* meters.

$L =$ __________ miles

g. Your capacitor would make a mighty large circuit element! How could it be made smaller physically and yet still have the same value of capacitance? You may want to examine the collection of sample capacitors for some ideas.

24.5. CAPACITORS IN SERIES AND PARALLEL

You can observe and measure the equivalent capacitance for series and parallel combinations. For this study you can use two identical capacitors. You'll need:

- 2 cylindrical capacitors, approx. 0.1 μF
- 1 capacitance meter

Recommended group size:	2	Interactive demo OK?:	N

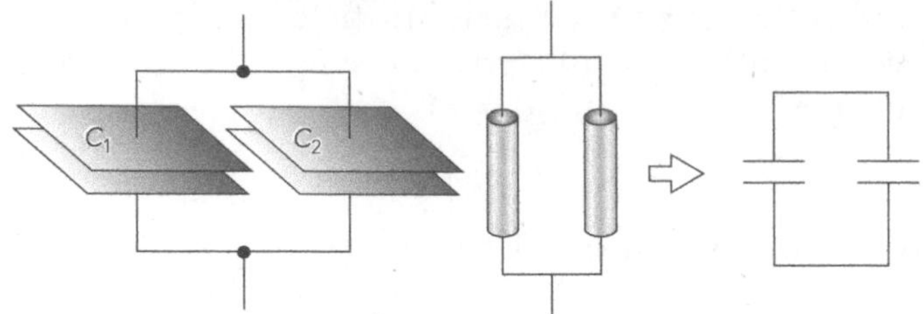

Fig. 24.5. Capacitors wired in parallel.

24.5.1. Activity: Capacitance for a Parallel Arrangement

a. Use direct physical reasoning to predict the equivalent capacitance of a pair of identical capacitors wired in parallel. Explain your reasoning below. **Hint:** What is the effective area of two parallel plate capacitors wired in parallel? Does the effective spacing between plates change?

b. What is the equivalent capacitance when your two pairs of aluminum sheets or your two cylindrical capacitors are wired in parallel? Summarize your *actual* data.

c. Guess a general equation for the equivalent capacitance of a parallel network as a function of the two capacitances C_1 and C_2.

$C_{eq} =$

Next, consider how capacitors that are wired in series, as shown in Figure 24.6 behave.

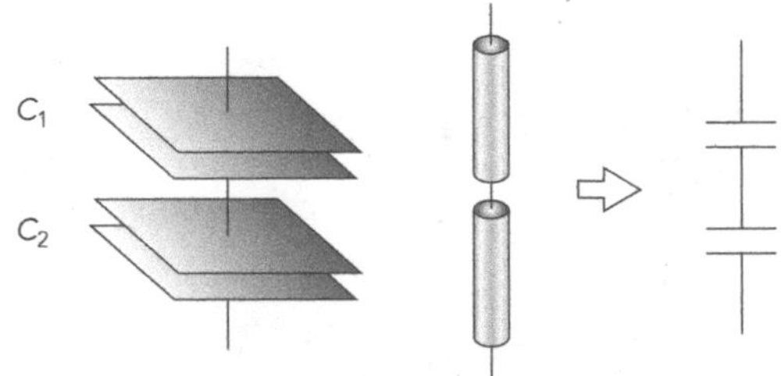

Fig. 24.6. Capacitors wired in series.

24.5.2. Activity: Capacitance for a Series Arrangement

a. Use direct physical reasoning to predict the equivalent capacitance of a pair of capacitors wired in series. Explain your reasoning. **Hint:** If you connect two capacitors in series, what will happen to the charges along the conductor between them? What will the effective separation of the "plates" be? Will the effective area change?

b. Measure the equivalent capacitance when your two pairs of aluminum sheets or your two cylindrical capacitors are wired in series. Report your actual data. Are the results compatible with the expected values?

c. Guess a general equation for the equivalent capacitance of a series network as a function of C_1 and C_2.

$C_{eq} =$

d. How do the mathematical relationships for series capacitors compare to those of resistors? Do series capacitors combine more like series resistors or parallel resistors? Explain.

CHARGE BUILDUP AND DECAY IN CAPACITORS

24.6. RC CIRCUITS

In the next section you will measure what happens to the voltage across a charged capacitor when it is placed in series with a resistor in a direct current circuit. Before making these measurements you should make some qualitative observations of capacitor behavior, so that you can explain what is happening as charge decays off a capacitor. For the observations in this section, you will need:

- 1 battery, 4.5 V
- 1 battery, 3.0 V
- 1 #14 bulb
- 1 #48 bulb
- 2 capacitors, .47 F
- 1 multimeter
- 6 alligator clip wires
- 1 SPST switch

Recommended group size:	2	Interactive demo OK?:	N

Qualitative Observations

By using a flashlight bulb as a resistor and one or more of the amazing new capacitors that have capacitances up to about a farad in a tiny container, you can "see" what happens to the current flowing through a resistor (the bulb) when a capacitor is charged by a battery and when it is discharged.

Fig. 24.7. A #14 bulb (rounded).

24.6.1. Activity: Capacitors, Batteries, and Bulbs

a. Connect a rounded #14 bulb in series with the 0.47 F capacitor, a switch, and the 4.5 V battery. Describe what happens when you close the switch. Draw a circuit diagram of your setup.

b. Now, can you make the bulb light again without the battery in the circuit? Mess around and see what happens. Describe your observations and draw a circuit diagram showing the setup when the bulb lights without a battery.

c. Draw a sketch of the approximate brightness of the bulb as a function of time when it is placed across a charged capacitor without the battery present. Let $t = 0$ when the bulb is first placed in the circuit with the charged capacitor. **Note:** Another way to examine the change in current is to wire an ammeter in series with the bulb.

d. Explain what is happening. Is there any evidence that charge is flowing between the "plates" of the capacitor as it is charged by the battery with the resistor (the bulb) in the circuit, or as it discharges through the resistor? Is there any evidence that charge is not flowing through the capacitor? **Hints:** (1) You may want to repeat the observations described in a. and b. several times; placing the voltmeter across the capacitor or placing an ammeter in series with the capacitor and bulb in the two circuits you have devised might aid you in your observations. (2) Theoretically, how should the voltage across the capacitor be related to the amount of the charge on each of its conductors at any given point in time?

e. What happens when *more capacitance* is put in the circuit? When *more resistance* is put in the circuit? (You can use a #48 bulb—the oblong one—in the circuit to get more resistance.) **Hint:** Be careful how you wire the extra capacitance and resistance in the circuit. Does more capacitance result when capacitors are wired in parallel or in series? How should you wire resistors to get more resistance?

Fig. 24.8. A #48 bulb (elongated).

A Capacitance Puzzle

Suppose two identical .47 F capacitors are hooked up to 3.0 V and 4.5 V batteries in two separate circuits. What would the final potential difference across them be if they were each unhooked from their batteries and hooked to each other without being discharged? This situation is shown in Figure 24.9. **Warning:** Be sure to hook the terminals that were charged positively together and those that were charged negatively together.

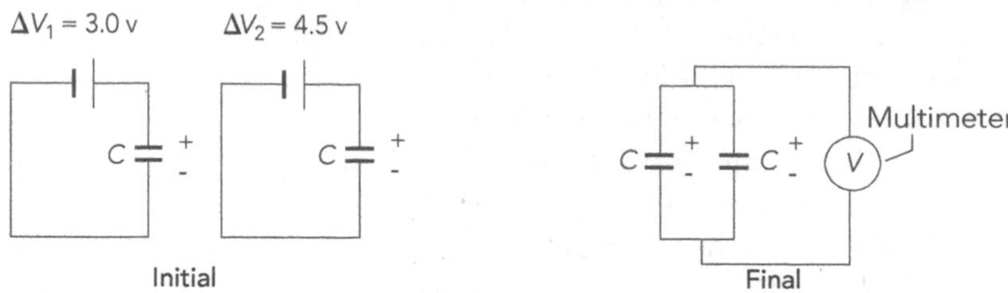

Fig. 24.9. A capacitor circuit. What is the multimeter reading ΔV_f?

24.6.2. Activity: Proof of the Puzzler

a. What do you predict will happen to the potential difference across the two capacitors? Why?

b. Can you use equations to calculate what might happen? **Hint:** What do you know about the initial charge on each capacitor? What do you know about the final sum of the charges on the two capacitors, if there is no discharge?

c. Set up the circuit and describe what actually happens.

d. How well did your prediction hold? Explain.

24.7. QUANTITATIVE MEASUREMENTS ON AN RC SYSTEM

The next task is to do a more quantitative study of your "RC" system. We will do this in two ways.

The first involves measuring the potential difference across a charged capacitor as a function of time when a carbon resistor has been placed in a circuit with it. A computer-based laboratory potential difference logging setup can be used to obtain data and view the trace of potential difference vs. time in graphical form as the capacitor discharges. The goal here is to figure out the mathematical relationship between potential difference across the capacitor and time that best describes the potential difference change as the capacitor discharges.

For the activities in this section you will need:

- 1 lantern battery, 6 V
- 1 capacitor, approx. 5000 μF
- 1 SPDT switch
- 1 resistor, 1.0 kΩ
- 1 computer data acquisition system
- 1 voltage measuring lead
- 1 electronic voltage probe
- 1 digital multimeter
- 6 alligator clip wires
- 1 capacitance meter (for $C \geq 5000\ \mu$F)

Recommended group size:	2	Interactive demo OK?:	N

A bulb is not a good constant value resistor because its resistance is temperature dependent and rises when it is heated up by current. For these more quantitative studies you should use a 1.0 kΩ resistor in place of the bulb while attempting to charge a 5000 μF capacitor. Wire up the circuit shown below in Figure 24.10 with a two-position switch in it. The switch will allow you to flip from a situation in which the battery is charging the capacitor rapidly to one in which the capacitor is allowed to discharge through the resistor more slowly. The voltmeter and leads to your computer interface should be placed in parallel with the capacitor—this allows you to measure the potential difference across it. (If you use an oscilloscope, set the time base control to 0.5 seconds per centimeter.)

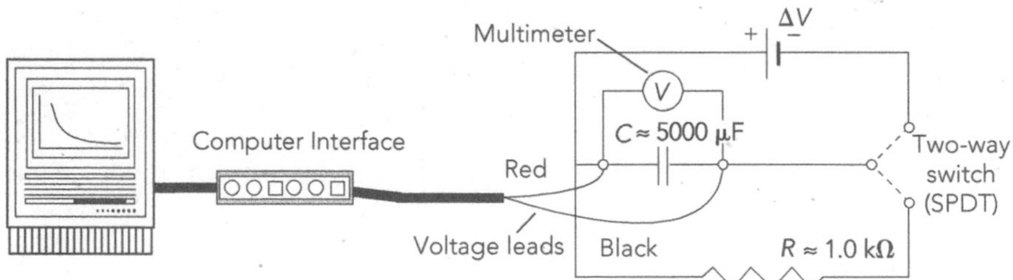

Fig. 24.10. RC circuit with potential difference measurements across a discharging capacitor.

24.7.1. Activity: The Decrease of Potential difference in an RC Circuit

a. Assume that the capacitor has been charged by the battery. What do you predict will happen to the potential difference across the capacitor when the two-position switch is flipped so that the battery is removed from the circuit? Explain the reasons for your prediction.

b. As soon as the switch is flipped from a battery terminal to a terminal of the capacitor, you can start measuring the potential difference across the capacitor at least every one or two seconds until the potential difference across the capacitor is about 1/100th of its initial value. Graph your data and guess how to model it. Affix (1) a table summarizing your data; (2) a graph of the V vs. t data and model in the following space. **Note:** Be sure to save your data on disk, as you will need to use it again in Activity 24.7.3. **Hint:** The ΔV vs. t curve cannot be modeled using a simple power or root of ΔV. Thus, you should try some simple functions to model ΔV such as: k/t, k, e^{-kt}, and so on where the k is a constant.

c. How did your observations fit with the prediction you made in part a?

The Theoretical RC Decay Curve

If you made careful measurements of ΔV vs. t for a capacitor, C, discharging through a resistor, R, you should have been able to plot what is known as an exponential decay curve. This curve has exactly the same mathematical form as the cooling curve you encountered in the study of heat and temperature. Mathematical reasoning based on the application of Ohm's law as well as the definition of current and capacitance can be used to predict exponential decay given by

$$\Delta V = \Delta V_0 e^{-\frac{t}{RC}} \qquad (24.2)$$

where ΔV_0 (typically equal to ΔV_B) is the initial potential difference across the capacitor (at time $t = 0$ s) and t represents the time elapsed since the switch was thrown to cut the battery out of the circuit.

24.7.2. Activity: Derivation of the Theoretical Decay Curve

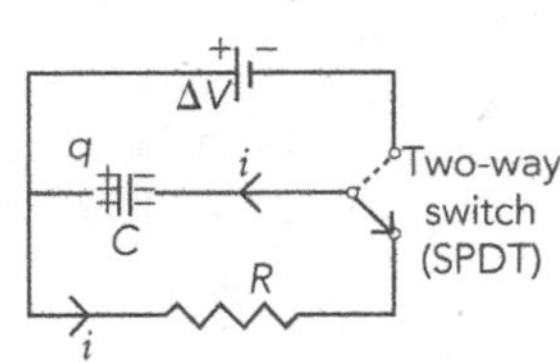

Fig. 24.11.

a. What is the equation for the potential difference difference across the capacitor shown in Figure 24.11. Express this in terms of the amount of excess charge $|q|$ on each capacitor plate and C? **Hint:** What is the definition of C?

b. What is the equation for the potential difference difference across the resistor in terms of the current, i, flowing through it (due to the discharge of the capacitor) and its resistance, R? **Hint:** Recall that potential difference *drops* in the direction of current flow.

c. Assume that the switch is in the *position shown in Figure 24.11* and that q represents the charge on the capacitor plate with excess positive charge. Show that at all times while the capacitor is discharging,

$$\frac{q}{C} = iR$$

Note: As the capacitor discharges, q gets smaller and smaller.

d. Using the definition of the instantaneous electric current passing *through* the resistor, explain why

$$i = -\frac{dq}{dt}$$

where q represents the excess charge on the positive capacitor plate (not the charge flowing through the resistor). **Hints:** (1) What is the source of the charge flowing through the resistor? (2) What is the relationship between the rate of flow of charge through the resistor, dq_R/dt and the rate at which excess charge flows off the capacitor plates, dq/dt?

e. Use the answers given previously to show that

$$\frac{dq}{dt} = -\frac{q}{RC}$$

in the circuit under consideration.

f. Show that the equation $q = q_0 e^{(-t/RC)}$, where q_0 is a constant representing the initial excess charge on the positive capacitor plate, satisfies the condition that

$$\frac{dq}{dt} = -\frac{q}{RC}$$

Hint: Take the derivative of q with respect to t and replace $q_0 e^{(-t/RC)}$ with q.

g. Use the definition of capacitance once again to show that theoretically we should expect that the potential difference across the capacitor will be given by

$$\Delta V = \Delta V_0 e^{(-t/RC)}$$

where ΔV_0 represents the initial potential difference across the capacitor.

Now comes the acid test: the comparison of the experimentally determined rate of the capacitor discharge to the theoretically predicted rate.

24.7.3. Activity: Does the Observed Decay Curve Fit Theory?

a. Carefully measure the R and C of the resistor and capacitor you used in the experiment in Activity 24.7.1 using a digital multimeter and a capacitance meter. List these values below (with proper units, of course). Also list the value of ΔV_0 from that experiment.

b. Develop a modeling worksheet. Set R, C, and ΔV as absolute parameters, using your measured values from part a, and calculate the theoretical ΔV vs. t for the circuit using the equations you derived in Activity 24.7.2. Create an overlay plot of the theoretical and experimental values for ΔV vs. t, using the experimental data you obtained in Activity 24.7.1. Affix a copy of your printout over the following sample worksheet.

	R(Ω)=	1000	Enter measured values, not these sample values.
	C(F)=	5.00E-03	
	ΔV_0(V)=	4.5	
RC decay curve			
ΔV (volts)			
t(s)	ΔV-exp	ΔV-theory	
0.0			
1.0			
2.0			
3.0			
4.0			
5.0			
6.0			
7.0			
8.0			
9.0			
10.0			
etc.			

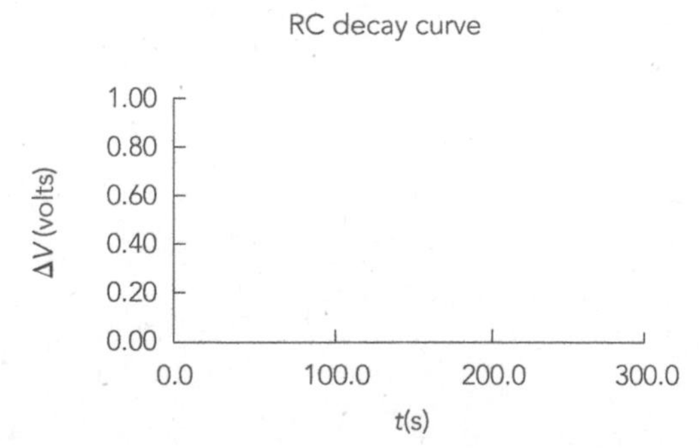

c. How well does theory match with experiment in this case?

d. What do you think would happen to the decay time if R were doubled? If C were doubled?

MORE ON CHARGE DECAY IN CAPACITORS

A Qualitative Summary of RC Decay

Let's consider the process of discharging a capacitor that is in series with a resistor one more time.

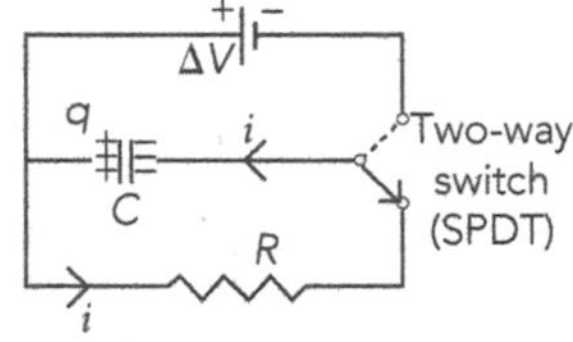

Fig. 24.11. Again!

24.7.4. Activity: Explaining Discharging Qualitatively

a. Assume that the capacitor is fully charged. When the switch is first flipped *so the battery is no longer in the circuit,* how much excess positive charge is on the positive capacitor plate? What is the amount of the potential difference ΔV_0 across the plates?

b. Is the current through the resistor a maximum or a minimum just after the switch is flipped? Does charge ever flow *through* the capacitor? Explain.

c. How is the potential difference across the resistor related to that across the capacitor?

d. What happens to the potential difference across the capacitor as charge drains away from it? Explain.

e. What happens to the potential difference across the resistor at the same time? Explain.

f. If the potential across the resistor starts to change, what must happen to the current in the circuit? Explain.

g. Why does the draining of charge from the capacitor eventually stop? Why does the current in the circuit go to zero?

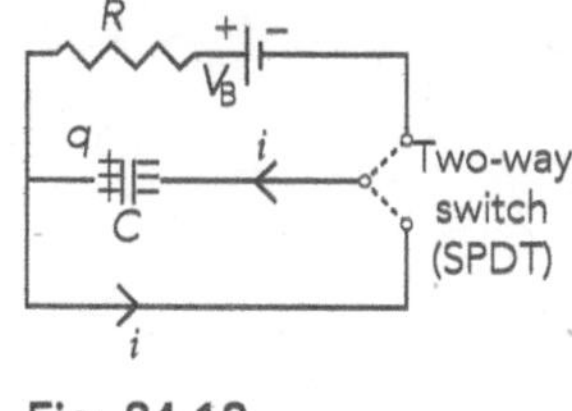

Fig. 24.12.

Capacitor Charging

If the resistor, R, in Figure 24.11 is moved up next to the battery as shown in Figure 24.12, an uncharged capacitor, C, can be charged by the battery in the presence of the resistor. The qualitative and quantitative considerations of this situation are very analogous to that of capacitor decay. For example, the capacitor charges up more rapidly at first when there are no charges on either of the capacitor plates to repel each other. Also, after a while when the potential difference across the capacitor is equal to that across the battery, the charging stops completely. It can be shown that the potential difference, ΔV, across the capacitor as it is charging is described by the equation

$$\Delta V = \Delta V_B(1 - e^{-t/RC}) \qquad (24.3)$$

where ΔV_B is the potential difference across the battery. This charging equation is used in the design of the flashing lights used at road construction sites and in the flash units used by photographers.

24.8. EXPONENTIAL DECAY: THEORY AND EXPERIMENT

During the last several activities you were asked to derive an equation that describes the decay of potential difference across a capacitor wired in series with a carbon resistor. Recall that the theoretical equation is given by

$$\Delta V = \Delta V_0 e^{(-t/RC)} \tag{24.4}$$

You were also asked to make measurements to verify the equation experimentally for only *one* resistor and capacitor combination. The exponential decay of potential difference across the capacitor is known as the *exponential relaxation of charge.* The product RC in a circuit has the units of a time and is called the "*RC time constant*" of the circuit because it determines the rate of decay of potential difference.

In the following activities, you are to conduct a more thorough theoretical and experimental investigation of the relationship between R, C, and the time it takes a discharging capacitor to "relax."

24.8.1. Activity: Theoretical RC Decay Times

a. Show that in a time equal to RC the potential difference ΔV across a capacitor drops to 36.8% of its initial value. **Hint:** Start with the capacitor decay equation and the definition of capacitance. Substitute RC for the time t and take the logarithm of both sides of the equation.

b. Another convenient equation is that describing the *half-life* of the capacitor decay as a function of the time constant, RC, of the circuit. Show that the half-life is given by the equation

$$t_{1/2} = RC \ln 2 = 0.693\ RC \tag{24.5}$$

Hint: Start with the capacitor decay equation. Substitute $V_0/2$ for the potential difference and $t_{1/2}$ for time. Take the logarithm of both sides of the equation.

You can use a computer-based laboratory system for logging the changes in potential difference as a capacitor discharges to verify the equation

$$t_{1/2} = RC \ln 2 = 0.693\ RC \qquad (24.5)$$

for several combinations of R and C. To do this you will need the following items:

- 1 multimeter (to measure resistances)
- 1 computer-based laboratory potential difference-logging system (with computer interface, test leads, data logger software)
- 1 battery, 4.5 V
- 1 capacitor, 5000 μF
- 5 assorted resistors, (such as 220 Ω, 470 Ω, 1.0 kΩ, 1.5 kΩ, 2.2 kΩ)
- 6 alligator clip wires
- 1 two-way switch
- 1 capacitance meter (> 5000 μF)

Recommended group size:	2	Interactive demo OK?:	N

You should set up a circuit similar to that shown in Figure 24.10 with the capacitor and one of the resistors. We could hold R constant and vary C and then hold C constant and vary R using the computer-based laboratory system. For this exercise, let's hold the capacitance in the circuit constant at 5000 μF, vary the resistance, and measure the resulting $t_{1/2}$. Thus, you should verify that

$$t_{1/2} = 0.693\ RC = [0.693\ C]\ R \qquad (24.5)$$

Some things to do first:

1. You need to calibrate the computer-based laboratory potential difference-logging system. To do this you should determine the potential difference of the battery that you plan to use with a multimeter and then follow the instructions on the computer screen when you have chosen the computer-based laboratory calibrate option.

2. You need to measure the value of the actual capacitor you plan to use with a good capacitance meter and substitute it into Equation 24.5 to get an equation of the form $t_{1/2}$ = (constant) R.

3. Figure out how to use the computer-based laboratory potential difference-logging system to find the half-life $t_{1/2}$ of the charge decay for the capacitor in an RC circuit.

24.8.2. Activity: An RC Decay Experiment

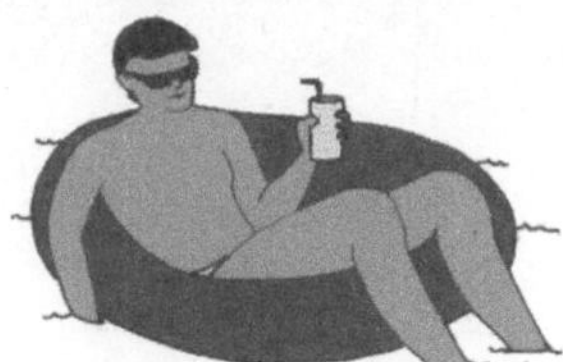

Fig. 24.13. The decay of charge from a capacitor in an RC circuit is sometimes called "relaxation."

a. Describe the method you are using to verify Equation 24.5. Include a sketch of your experimental setup and a circuit diagram in the space below.

b. In order to compare theoretical values of $t_{1/2}$ with experimental values, you should measure the C value of your capacitor several different times and list its average and standard deviation, with units, in the space below. Also list the value of $0.693C$ and its standard deviation. Don't forget to include your units.

Trial	C	Units
1		
2		
3		

$C_{avg} =$ ________ ± ________

$0.693\ C_{avg} =$ ________ ± ________

c. Perform the experiment (in which you determine the half-lives of the RC decay with C *approximately* equal to 5000 μF) using *at least five different* resistance values. After some practice, create a data table of half-lives for each of your measured R values (complete with correct units!) in the space below. **Warning:** Use *measured,* not rated, values of resistance.

d. Print out a graph showing the decay curve (ΔV vs. t) for at least one of your RC combinations and annotate it to show how you determine the value of $t_{1/2}$ using the "Analyze" feature of the data logger software. Affix the graph in the space below.

e. Finally, enter your $t_{1/2}$ and R data into a spreadsheet and print out a graph of $t_{1/2}$ vs. R and affix it in the space below.

f. According to Equation 24.5, what is the expected value of the slope of the $t_{1/2}$ vs. R graph shown in part e, if the 5000 μF rating of the capacitor is the actual value of the capacitance?

g. Discuss the data and graph you reported in parts c. and e. How linear is the graph? How close is the slope to the expected value? If your experimental and theoretical slopes differ, what do you think is the "best value" for the capacitance of the "5000 μF" capacitor?

Name ______________________ Section __________ Date __________

UNIT 25: ELECTRONICS

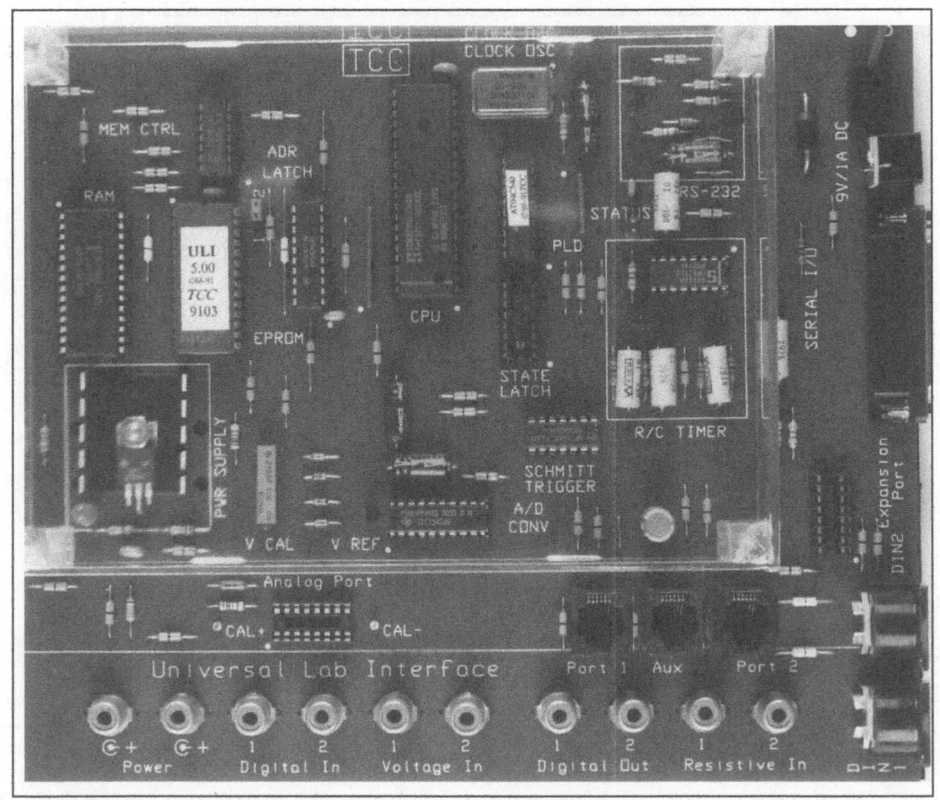

This photograph shows an early model of a printed circuit board and electrical components that made up the Universal Laboratory Interface originally designed for use in the Workshop Physics program. In addition to the familiar looking resistors and capacitors on the board, there are long black plastic elements known as integrated circuits (or ICs for short). These integrated circuits each contain thousands of tiny circuit elements including semiconducting devices known as transistors as well as resistors and capacitors. In this unit you will learn to use a popular integrated circuit to amplify sound, and other common integrated circuits to construct a digital stopwatch. An oscilloscope will be used to observe changes in voltage between points of interest in your circuits.

UNIT 25: ELECTRONICS

First there was a generation of huge machines that relied on vacuum tubes; then there was one that used transistors; a third relied on the integrated circuits printed on silicon chips; and a fourth—a generation only now emerging—turns to very large scale integrated circuits (VLSI), with chips so compact they must be designed by another computer. The achievement of artificial intelligence would be the crowning effort, the fifth generation.

Richard Flaste

OBJECTIVES

1. To understand how an oscilloscope can be used to display changes in potential difference over time.
2. To investigate different direct current (dc) and alternating current (ac) wave forms.
3. To gain experience with fundamental logic operations used in digital electronics.
4. To learn about the functional characteristics of certain popular integrated circuits such as op amps, counters, and digital logic elements.
5. To construct a circuit designed to carry out a useful operation—that of a stopwatch.

25.1. OVERVIEW

Electronics is the sub-field of electricity that deals with the information contained in electrical signals, such as those that produce sound from a loudspeaker, a TV picture, or memory states in a computer.

This unit is intended to provide you with a brief introduction to electronics, including some of the devices used in circuits and some of the ways you can transform electrical signals by designing and constructing circuits. In the first part of this unit you will learn to use an oscilloscope, which is one of the most basic measuring instruments used in electronics to measure potential difference changes. Next you will explore analog electronics by constructing a simple amplifier to boost a weak electrical signal so that it becomes an audible sound when attached to a loudspeaker. Then you will begin an exploration of some of the digital electronic components that provide the basis for the modern digital computer. Finally you will use digital electronic components to build a stopwatch. The projects in this unit might stimulate you to learn more about electronics on your own or to take a course in electronics.

THE OSCILLOSCOPE

25.2. INTRODUCTION TO THE OSCILLOSCOPE

Many scientists and engineers are interested in measuring potential differences and visualizing how they change over time. For example, in the next few activities you will be studying the changes over time of the potential difference coming from the earphone output of a cassette recorder. Although these kinds of potential difference measurements can also be made using newly developed computer systems, the oscilloscope is presently the most commonly used instrument for making such measurements.

The heart of an oscilloscope is an evacuated glass tube known as a cathode ray tube, or CRT for short. It contains an electron gun and two pairs of parallel metal plates that can hold electrical charge on them. When two parallel plates carry charges of equal magnitude but opposite sign, the magnitude of the charge on the plates is proportional to the "potential difference" between them. As you have just learned, two charged plates will produce an electric field between them.

By producing a beam of electrons and then deflecting it under the influence of the electric field that results from the potential difference used to charge the plates, we can measure the potential difference.

A schematic of the CRT is shown as follows.

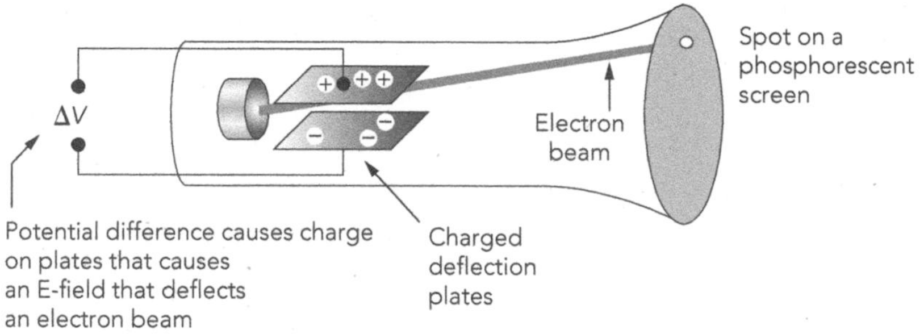

Fig. 25.1. A simplified view of the tube in an oscilloscope.

It turns out that the motion of an electron moving at an initial speed v perpendicular to a uniform electric field between two charged plates will undergo the same type of parabolic motion that a projectile will if it is shot horizontally off a cliff in a gravitational field.

25.3. MOTION OF AN ELECTRON IN A UNIFORM E-FIELD

It is possible to prove mathematically that the deflection of an electron passing between charged metal plates is proportional to the potential difference across the plates. It is this proportional relationship between potential difference and deflection that allows us to use an oscilloscope to display potential difference changes graphically.

Let's explore the relationship between potential difference and deflection in more detail. Except near the edges of the plates, a pair of parallel plates with an amount of potential difference $|\Delta V|$ across them will have a uniform electric field between them of magnitude $E = |\vec{E}|$ given by

$$E = \frac{|\Delta V|}{d} \tag{25.1}$$

where d is the spacing between the plates. The direction of the field will, of course, be determined by a vector pointing away from the positively charged plate and toward the negatively charged plate. A cross section of parallel plates is shown in Figure 25.2.

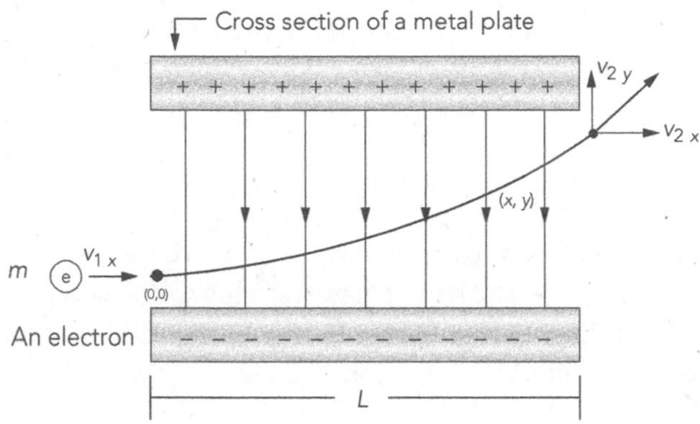

Fig. 25.2. Oscilloscope deflection plates having a potential difference across them deflect an electron with an initial velocity of $v_{1\,x}$ to a final velocity with components $v_{2\,x}$ and $v_{2\,y}$.

As background for understanding the oscilloscope, let's describe the motion of an electron as it passes through an electric field between two parallel plates. **Note:** To avoid confusion in the mathematical description, we will use a capital V when referring to the potential difference ΔV between the plates and describe the velocity components of the electron with a lowercase v.

25.3.1. Activity: Proof that Deflection is Proportional to Potential difference

a. An electron of charge $q_e = -e$ is placed in an electric field $\vec{E}$. What is the expression for the force $\vec{F}$ on the electron in terms of the charge on the electron, q_e, and the electric field $\vec{E}$ across the plates?

b. If the mass of the electron is denoted by m, what is the mathematical expression for its acceleration, $\vec{a}$, in terms of q_e, m, ΔV, and d? If the electric field between the plates is uniform—that is, a constant, what can we say about the acceleration of an electron as it passes between the plates?

c. Assume that the uniform electric field between the plates is in the y-direction. Show that if the initial velocity, $v_{1\,x}$, is in the x-direction, it remains constant as the electron moves perpendicular to the electric field so that $v_x = v_{1\,x}$.

d. If the plates have a length L, what is the time it takes an electron moving at an initial velocity $v_{1\,x}$ to pass between them?

e. What is the velocity component, $v_{2\,y}$, of the electron in the y-direction at a time $t = t_2$ after it has traveled a distance L in the x-direction? **Hint:** Express your result as a function of a and t and then as a function of q_e, m, ΔV, d, $v_{1\,x}$, and L.

f. Is $v_{2\,y}$ proportional to $|\Delta V|$? Why?

g. Suppose the electron leaves the influence of the electric field and travels an *additional* distance of D and hits the face of the phosphorescent screen. Explain why the fact that $v_{2\,y}$ is proportional to V means that the displacement, y, of a spot on a screen is approximately proportional to the potential difference, V, between the plates.

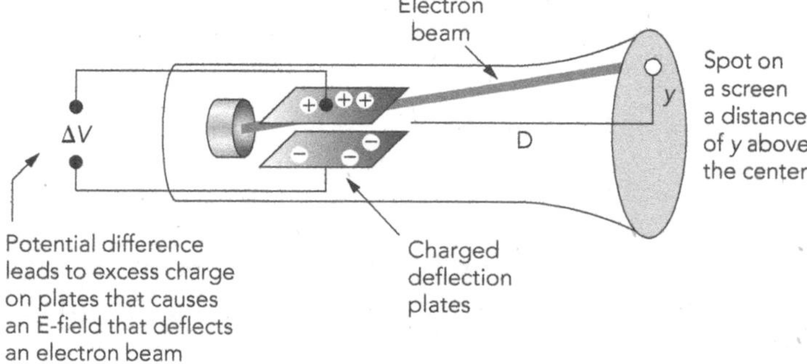

Fig. 25.3. Deflection of electrons in an oscilloscope.

25.4. PRODUCING POTENTIAL DIFFERENCE CHANGES FOR INPUT TO A SCOPE

In order to observe how an oscilloscope functions, you will need to put potential differences that change over time into it and see what happens. The following equipment is recommended:

- 1 function generator with clip leads
- 1 small speaker
- 1 BNC to clip lead cable

Recommended group size:	3	Interactive demo OK?:	N

Using a Function Generator

You will use a function generator similar to that pictured below to produce potential differences that change with time in a regular pattern. If these potential difference patterns are put into a speaker, you will hear sounds. By turning the dial and pushing the "Frequency Multiplier" buttons, you can change the basic pitch of that sound wave. By choosing a different "function" (sine wave for a pure tone, triangle wave or square wave for "distorted tones") the timbre of the sound coming out of the speaker is changed.

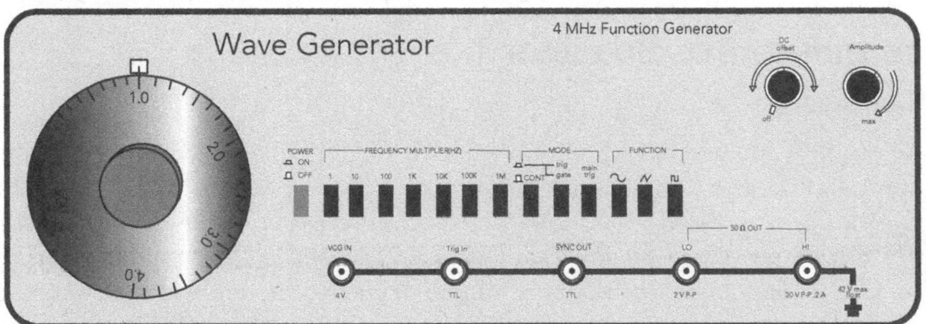

Fig. 25.4. A typical function generator.

To begin, let's set up the function generator to produce sounds:

1. Plug in and turn on the function generator. Turn off the DC OFFSET.
2. Connect the cable with the clip leads to the LOW potential difference output.
3. Set the function generator to produce a 256-Hz sinusoidal wave by setting the frequency dial to 2.56 and the frequency multiplier to 100.
4. Attach the two output clip leads from the function generator across a speaker and adjust the AMPLITUDE dial until you can hear a tone. **Note:** If you can't hear the tone, then try the high potential difference output.

Warning! Do not blast your ears or the speaker with large amplitude waves or with very low or very high frequencies (even at small amplitudes). These practices can damage the speaker and/or eardrums.

25.4.1. Activity: Sounds from a Function Generator

a. Describe the sound you hear when you have the sine wave output set at 256 Hertz.

b. Describe what happens to the sounds when you use the triangle and square wave outputs instead.

c. What happens to the sounds when you change the frequency of the function generator output?

d. How do changes in amplitude affect the sounds?

25.5. USING AN OSCILLOSCOPE

More Theory

The oscilloscope uses a cathode ray tube or CRT to display the effects of two potential differences at the same time. *Usually, one potential difference is the potential difference in the circuit that is of interest, and the other potential difference is provided by the oscilloscope itself and is proportional to time.* The CRT thus acts to plot the potential difference of interest against time. In a sense, the oscilloscope represents "electronic graph paper."

The CRT shown in Figure 25.5 consists of an electron gun that uses a potential difference of ΔV_A to accelerate electrons to a velocity, v_{1x}, parallel to the axis of the tube. It has two pairs of deflection plates—one pair that can give an electron a vertical (up or down) deflection and one pair that can give an electron a horizontal (side-to-side) deflection. Note that only the plates causing vertical deflections are shown in the diagram. The tube has a phosphorescent surface on its "face" that glows when electrons hit it.

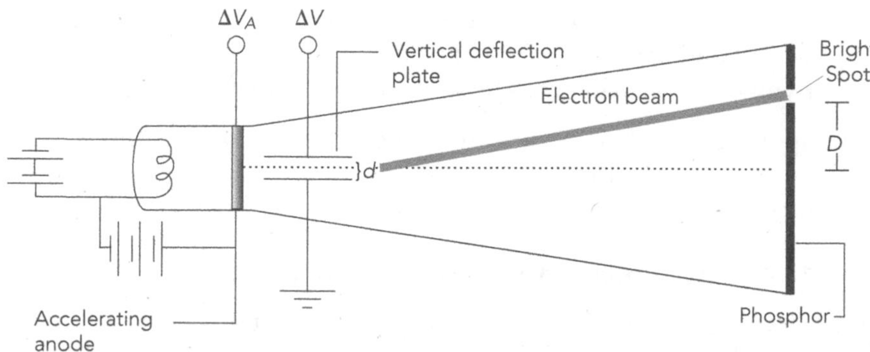

Fig. 25.5. Drawing of a cathode ray tube.

The inside of the tube is covered with a conductive powder called aquadag, and this provides a "return" path to complete a circuit inside the tube. The electron gun consists of a charged heated wire with excess negative charge on it and a plate with a hole in it called the anode with excess positive charge on it. The heated wire is much like the filament in a light bulb. It is the source of the electrons that are literally boiled away from it and repelled by the heater's excess negative charge toward the anode's positive charge. Most of these accelerated electrons fire out the hole in the anode, which is the "muzzle" of the electron gun.

A grid of fine wires is placed between the cathode and the anode; these wires can modify the potential difference that the electrons leaving the cathode "see." We have left this grid out of our diagram. It can be given a negative potential large enough to keep most of the electrons boiled off from the cathode from being accelerated by the anode and reaching the phosphor, and thus serves as a brightness control.

In order to create a graph of potential difference changes over time, a second pair of plates at right angles to those pictured are included to provide a *sweep potential difference.* Internal circuitry in the oscilloscope is designed to cause a beam to "sweep" horizontally from left to right across the screen at a rate that can be set. The plates causing the sweep are given a potential that varies in time in a saw-tooth pattern, as shown.

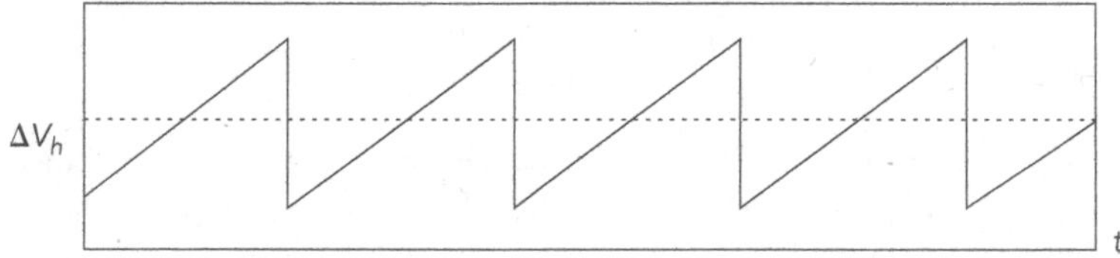

Fig. 25.6. Saw-tooth potential difference causing the horizontal sweep.

The net result of all this is to make the spot where the electrons strike the phosphor move smoothly from left to right at a fixed rate and then reappear at the left edge again for another sweep across. The speed with which the electrons move along is called the time base, and it is adjustable.

In order to learn to use an oscilloscope to measure constant and changing potential differences, you will need the following equipment:

- 1 oscilloscope (with a BNC to clip lead cable)
- 1 demonstration oscilloscope (optional)
- 1 battery, 5 V
- 1 function generator (with a BNC to clip lead cable)

Recommended group size:	3	Interactive demo OK?:	N

The face of an oscilloscope is shown in Figure 25.7.

Using Some of the Controls

1. Plug in the oscilloscope at your lab station.
2. Turn the CRT-Intensity control (directly to the right of the tube face and near the top) almost fully counterclockwise.

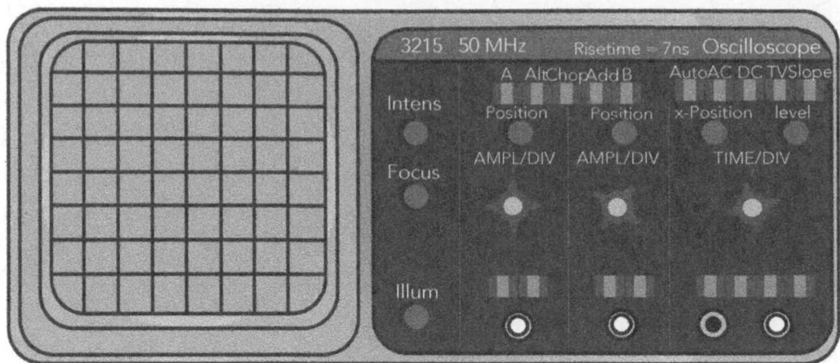

Fig. 25.7. An oscilloscope face.

Warning: Do not leave the intensity control so high that a halo appears. Such high intensity is bad for the screen.

3. Turn the Power/Illumination control clockwise to "on."
4. Push the *A Channel* and AUTO buttons as shown in Fig. 25.7.
5. Let the scope warm up a bit.

25.5.1. Activity: The Oscilloscope Controls

a. Play with the intensity control and the power/illumination controls. What do they do?

b. Use the intensity control to adjust the brightness of the spot on the oscilloscope screen so that it is just comfortably visible. Play with the Focus control. What does it do?

c. With the auto button depressed, set the time base knob (Time/Div) to 0.5 seconds per centimeter so that the beam of electrons sweeps horizontally across the phosphor at the rate of one-half second for each box on the grid. All other push buttons should be in the normal (out) position. You should see a trace moving across the screen. Play with the *time base* control. What does it do? What do the *position* controls do?

d. Next, use the oscilloscope to measure the steady potential difference from a battery. Attach one of the "BNC" to clip lead cables in the lab to the "A" input and connect the battery across the "A" input to the scope. Set the AC/DC coupling to DC. Next set the "A" sensitivity control to 2.0 volts per division and then experiment with a variety of different sensitivity or AMPL/DIV settings. What happens? What do the sensitivity controls do?

Measuring Wave Forms

Set up the function generator to produce a 256-Hz sinusoidal wave form, just as you did earlier, and examine this with your oscilloscope. Experiment with changing the amplitude of the wave form by adjusting the amplitude control on the function generator. Experiment with putting in triangle and square waves from the generator to the oscilloscope.

25.5.2. Activity: Measuring Changing Potential Differences

a. With the function generator set at 256 Hz, use your oscilloscope to determine the period of a triangle wave form. **Hint:** Make sure the inner knob of the time base control is in the calibrate position and then read the time base setting.

b. How does this period compare with the period calculated on the basis of the frequency reading on the dial?

c. Experiment with the "DC Offset" control on the function generator and the "AC/DC" button on the oscilloscope. Explain how these controls affect the oscilloscope display.

d. Switch over to the sinusoidal wave and square wave outputs of the function generator. Sketch each wave form in the space below.

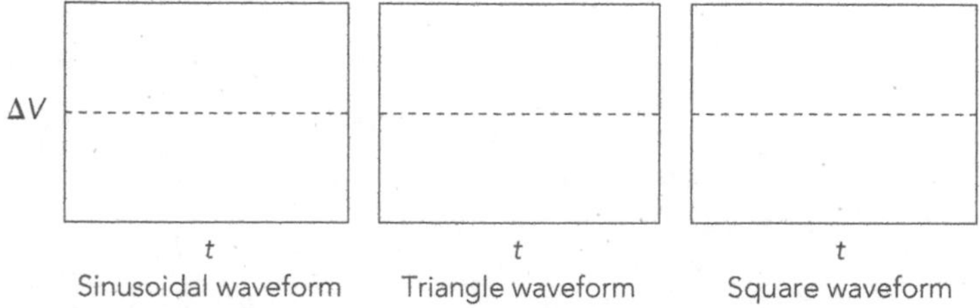

Sinusoidal waveform   Triangle waveform   Square waveform

e. Experiment with the frequency dial and multipliers on the function generator and the time base control on the oscilloscope. How do changes in these settings affect the wave forms?

25.6. ELECTRIC POWER

As a battery or another potential difference source pushes electric charges through a circuit, these charges lose potential energy. Consider a simple direct current circuit in which there is a potential difference V across a resistor R. How fast does the charge flowing through the resistor lose energy? What's its power loss? Let's observe what happens to the resistor as charges pass through it. Then let's do some simple mathematical reasoning to predict what happens quantitatively. For this next activity you will need the following items:

- 1 battery, 4.5 V
- 1 protoboard
- 3 resistors (such as 10 Ω, 22 Ω, and 39 Ω)
- 4 alligator clip leads
- 1 micro switch
- 1 digital multimeter (to measure ohms)

Recommended group size:	4	Interactive demo OK?:	N

25.6.1. Activity: Power Loss in a Simple Circuit

a. Since you will be setting up a circuit with a small micro switch, you should play with one to get a feel for what it does. To do this you can use the ohmmeter feature of a multimeter to determine the resistances between the switch terminals. **Note:** If a "1" appears at the left of the multimeter display, this indicates that the resistance is too large to measure with the currently selected scale.

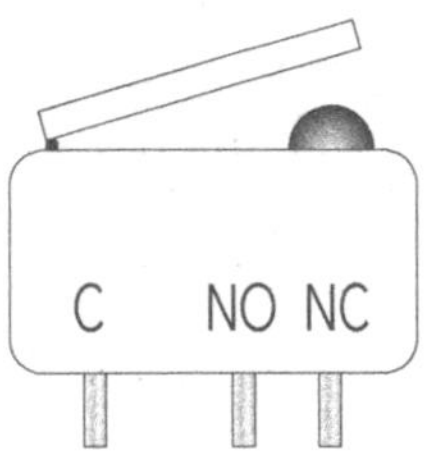

When the switch is not depressed, what is the resistance between the common terminal (marked with a C) and

1. The normally open (NO) terminal? ____
2. The normally closed (NC) terminal?____

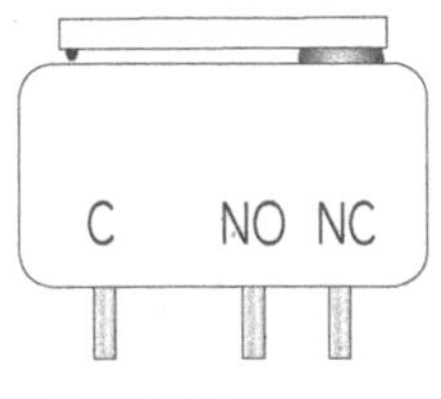

Fig. 25.8.

When the switch is depressed, what is the resistance between the common terminal (marked with a C) and

1. The normally open (NO) terminal?_____
2. The normally closed (NC) terminal?____

b. Set up a series circuit with a battery, resistor, and micro switch. Close the switch and feel the resistor as current flows through the resistor. What happens? Into what form has the electrical energy been transformed?

c. Does the effect you noticed in part b. become more pronounced or less pronounced when the resistor has a larger value? Is this what you expected? Why or why not?

d. Suppose that in a time interval Δt an amount of charge $|q|$ loses $|q||\Delta V|$ of energy in passing through a resistor R and undergoing a potential drop of ΔV. What is the equation for the current as a function of Δq and Δt?

e. Show that the loss of power, P, when current undergoes a potential drop of ΔV through a resistor R in passing is given by $P = i\Delta V$.

f. Use Ohm's law and the equation in part e. above to show that $P = (\Delta V)^2/R = i^2 R$ where I is the current is passing through a resistance of R.

g. Which equation in f. above will help you describe the qualitative observation you made in part b.? Is the equation consistent with your qualitative observation? Explain.

ANALOG AND DIGITAL ELECTRONICS

25.7. INVESTIGATING A "LOW LEVEL" TAPE RECORDER SIGNAL

Suppose you want to hook up the output of a cassette tape deck to a speaker and an oscilloscope to listen to and look at musical tones. Since tape decks are expensive, your instructor may substitute a less expensive cassette recorder and ask to you hook up a special earphone cable (with a 10 KΩ resistor in series with it) to the cassette recorder. This will give the recorder the electrical characteristics of a tape deck.

For this activity you will need the following items:

- 1 oscilloscope
- 1 speaker, 8 Ω
- 1 BNC to clip lead cables

- 1 cassette recorder with output (or cassette tape deck)
- 1 special earphone output cable with a 10-KΩ series resistor
- 1 audio tape with musical tones on it
- 1 digital multimeter (to check circuit if needed)

Recommended group size:	4	Interactive demo OK?:	N

You should connect the cassette recorder to the speaker and oscilloscope as shown in Figure 25.9.

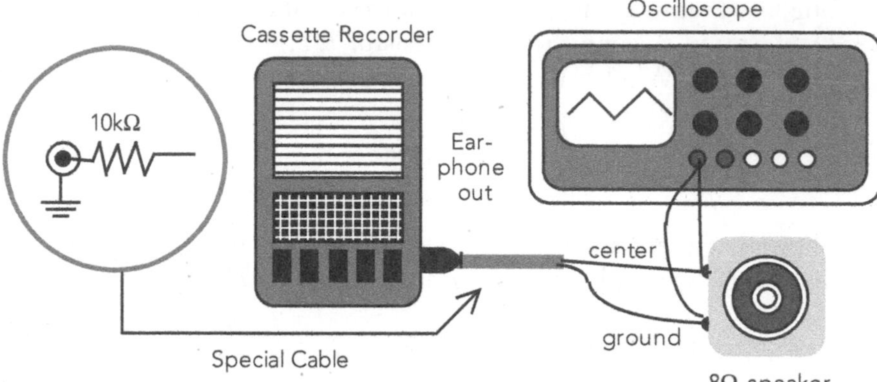

Fig. 25.9. Setup for the display of an audio signal showing the special earphone jack cable with a 10 KΩ resistor wired in series so that the recorder has a high impedence output like that of a tape deck.

25.7.1. Activity: Looking at and Listening to Signals

a. Assume that the earphone output of the cassette recorder acts like a battery with an internal resistance r_i and that the speaker and oscilloscope inputs have resistances denoted by $R_{speaker}$ and R_{scope}, respectively. Examine your hookup wiring and draw a circuit diagram for the setup. **Note:** The analogy between the tape recorder output and a battery is not really very good since the potential difference output stimulated by the audio information stored on the tape causes the output potential difference of the cassette recorder to vary in time.

b. Describe and sketch the trace of a typical signal on the oscilloscope without the loudspeaker hooked up.

c. What happens when a loudspeaker is connected across the output of the cassette recorder? Do you hear any sound from the loudspeaker? If not, what might the problem be?

25.8. ELECTRICAL ENERGY AND POWER

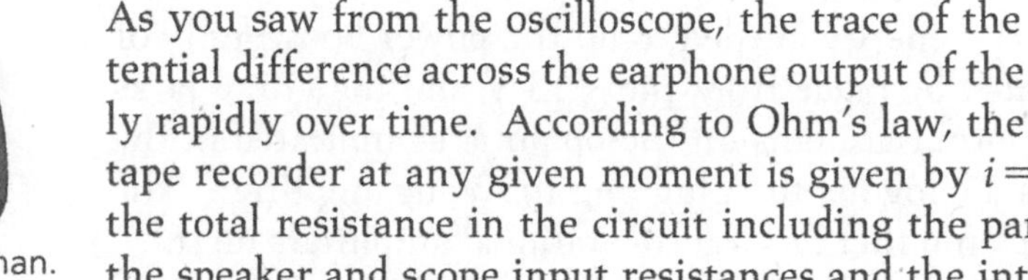

Fig. 25.10. A Walkman.

As you saw from the oscilloscope, the trace of the potential difference or potential difference across the earphone output of the tape recorder changes fairly rapidly over time. According to Ohm's law, the current flowing out of the tape recorder at any given moment is given by $i = \Delta V/R$ where R represents the total resistance in the circuit including the parallel network made up by the speaker and scope input resistances and the internal resistance of the cassette recorder output. This is shown schematically in Figure 25.11.

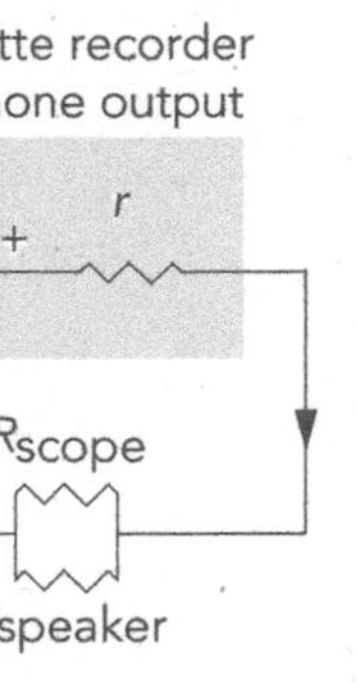

Fig. 25.11. Circuit showing input resistances.

It turns out that the input resistance of an oscilloscope is much, much larger than that of a speaker, and for our purposes essentially all of the current, I, that is put out by the cassette recorder flows through the speaker. However, in driving a speaker, a certain amount of electrical power is needed. Power, by definition, is the rate at which the electrical charges that flow through the speaker lose energy. In the last unit you derived the equation for power loss through a resistor as $P = VI$. Thus, there are two ways to boost the power delivered to the speaker. One is to increase the potential difference output of the cassette recorder and the other is to increase the current that is put out by lowering the effective internal resistance of the potential difference source and hence boosting the current coming into the speaker. The operational amplifier that you will use below to obtain more power is actually a current booster.

25.9. BUILDING AN AMPLIFIER TO BOOST CURRENT FROM AN AUDIO TAPE PLAYER

Let us now construct an amplifier to take the low power signal from the cassette recorder and reproduce the wave shape with greater power. Adjust the oscilloscope to observe the low-power electrical output signal from the audio cassette recorder when the cassette is playing a tape.

For this activity you will need the following items in addition to the oscilloscope and speaker you used in the last activity:

- 1 oscilloscope
- 1 speaker, 8 Ω
- 1 BNC to clip lead cable
- 1 cassette recorder with earphone output (or cassette tape deck)
- 1 special earphone output cable (10 kΩ series resistor built into plug)

- 1 audio tape with musical tones on it
- 1 LM324 op amp (an alternate number is ECG987)
- 1 powered protoboard (with ± 15V)
  (or a small protoboard and a collection of batteries)
- assortment of small lengths of # 22 wire (for use with the protoboard)
- 1 digital multimeter (to check circuit if needed)

Recommended group size:	4	Interactive demo OK?:	N

The LM324 integrated circuit is an op amp (operational amplifier) that is designed to amplify low power signals. *As is the case with all the integrated circuit elements to be used in this unit, the op amp is an active circuit element, and cannot operate unless a supply potential difference* (that is, power) *is applied to it.* (This is a matter of conservation of energy since power can't be boosted without adding the energy needed to do the power boosting.) For this op amp, connections must be made from the +15 V and the -15 V jacks on the protoboard to the appropriate pins on the op amp, as indicated in the pin diagram below. This is analogous to "plugging in" your amplifier. The LM324 actually is a quad op amp that has four operational amplifiers on it.

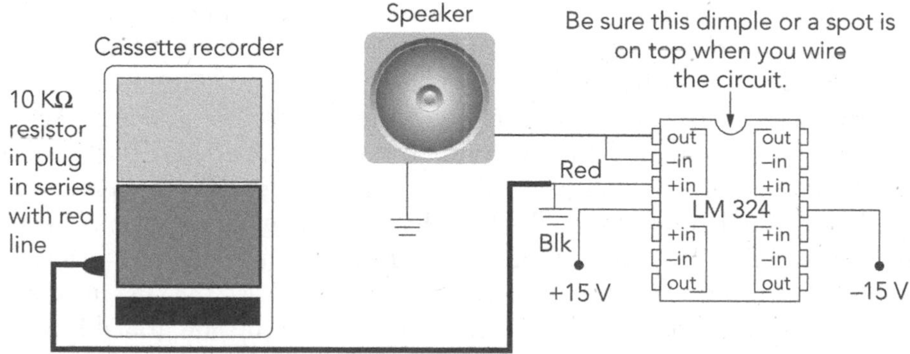

Fig. 25.12. Schematic for attaching a cassette recorder to an amplifier. The outer sheath of the special earphone cable is connected to ground (0 V) and the center (red) conductor is connected in series with a 10 KΩ resistor and then to the input to the LM324 op amp.

About Integrated Circuits

Integrated Circuits, or ICs, are amazing devices in which tiny transistors (which you haven't learned about yet), diodes, capacitors, and resistors are connected together with thin metal films to make elaborate circuits. The techniques for making ICs have improved so much in the past few years that we have advanced from having ICs to VLSIs (very large-scale integrated circuits). VLSIs have up to a million circuit elements in them and can do very complex things. When an IC is bundled together and put in a single package, it is often encased in black plastic that has two rows of connectors on it. Thus, an IC tends to look for all the world like a *bug* in the centipede or millipede family. You will be using several types of "bugs" in this unit.

The diagram in Figure 25.12 is for a circuit you are about to wire up. A new symbol appears in two places on the diagram and looks like this: ⏚. This symbol represents a concept called ground. An electrical ground is a common tie point for many elements in a circuit. (This common tie point is often chosen to have a large capacity for holding charge. A system. of metal water pipes makes a good ground.) **Notes:** (1) It is important to realize that all the points in a circuit diagram that are marked with a ground symbol, ⏚, should be connected to each other. (2) See the Appendix on the last page of this unit for protoboard wiring conventions when using ICs like the LM 324.

25.9.1. Activity: Seeing and Hearing Amplified Signals

a. Connect the output from the cassette recorder to the "+in" input of one of the op amp circuits. Connect the output of the amplifier to the loudspeaker and connect the other lead from the loudspeaker connected to ground), and describe the sounds you hear at various places along the tape.

b. Connect the oscilloscope to the amplifier output. Describe the oscilloscope potential difference vs. the patterns observed and how these patterns are related to the oscilloscope potential difference patterns you hear. Identify the frequency of the tone at two places along the tape. Explain how you determined these frequencies.

25.10. DIGITAL ELECTRONICS

The amplifier you just built was an analog amplifier. The term "analog amplifier" refers to a circuit that can take an input potential difference that can vary continuously over time and either boost the potential difference or current to create a higher power output signal that also varies continuously over time. Many ICs are digital rather than analog circuits. Let's explore the meaning of the term "digital."

Counting and displaying numerals, as in a digital watch, are operations that require a different type of electronic signal and circuitry than the continuously varying signal characteristic of audio electronics. *Digital electronics* involves potential differences that are either "on" or "off." The potential difference is either zero ("off") or at some other fixed potential difference which is defined as "on." The 0 and +5 Volt digital circuitry you will be using will interpret any potential difference greater than about 3 V as an "on" state and any potential difference less than about 2 V as an "off" state. One advantage of digital circuitry is that it is remarkably insensitive to stray variations or "noise." Thus, digital recorders, computers, and other modern digital devices are vastly more accurate and reliable than circuitry that attempts to reproduce varying signals.

We will not try to understand the basic physics behind the operation of the digital circuits in this unit. Instead, you will treat each digital "bug" as a "black box" and attempt to understand the logic that determines the nature of its output potential differences as a function of its digital input potential differences. In other words, we are interested in how various digital circuit elements behave (rather than analyzing why they have that behavior).

The inputs and resulting outputs for a digital circuit element can be summarized in something called a *truth table.*

The Inverter and the AND Gate

First you are going to explore the outputs of two very basic digital ICs as a function of their inputs. This will help you prepare to undertake a more extensive digital electronics project—the building of a digital stopwatch.

For the next few activities you will need the following items:

- 1 Inverter, 4049
- 1 AND gate, 4081
- 1 LED
- 1 powered protoboard (with +5 V) (or a small protoboard and a collection of batteries)
- assorted small lengths of wire

Recommended group size:	2	Interactive demo OK?:	N

(An LED, or *light emitting diode,* is a simple device that can be used to indicate whether a potential difference is "on" or "off.") Starting with a 4049, with the pin designations as shown (don't forget to supply power to this active circuit element!), connect the LED to the output of one of the six identical independent circuits. (The flat side of the base of the LED bulb should be connected to ground.)

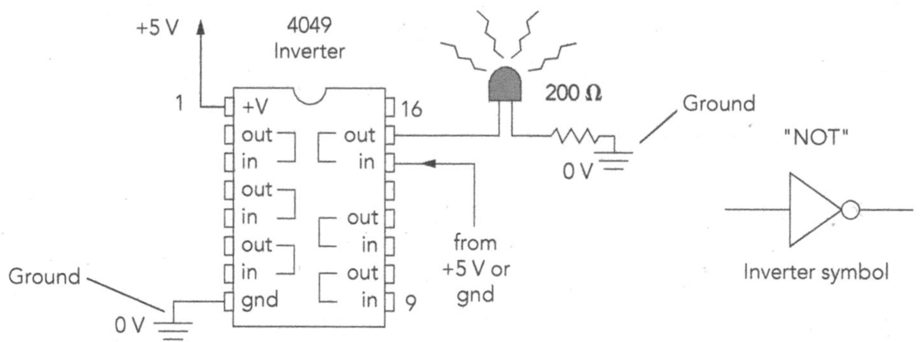

Fig. 25.13. Wiring an LED to an inverter.

25.10.1. Activity: Truth Table for the 4049 Inverter

a. Check the output when the corresponding input is zero or 5 V. Complete the "truth table" shown below, in which "0" represents zero volts (off) and "1" represents a potential difference that is close to five volts (on).

Truth table	
Input	Output
0	
1	

b. Why is this circuit called an "inverter"?

The 4081 "AND" bug contains four identical AND gates. Each gate has two inputs and one output. Wire up the output of one of the gates to the LED, which is in turn connected to ground.

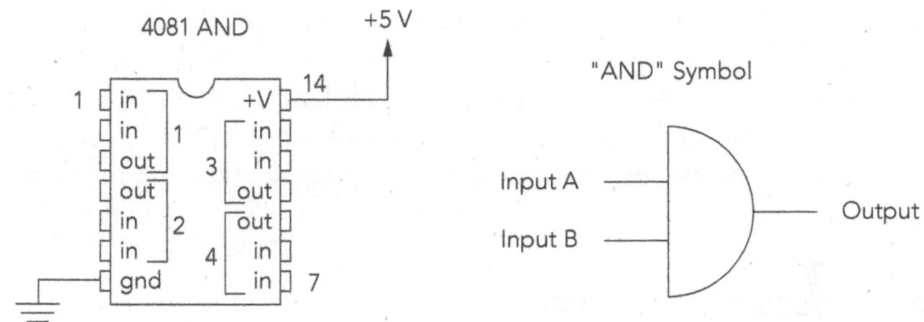

Fig. 25.14. Pinouts for a 4081 AND integrated circuit.

25.10.2. Activity: Truth Table for the 4081 AND Gate

a. Check the output for each combination of inputs and complete the "truth table" shown below.

Input A	Input B	Output
0	0	
0	1	
1	0	
1	1	

b. Examine the truth table carefully. Predict what the output would be if input A is held at 0, and input B is a continuous string of pulses – that is, the potential difference alternates between 0 V and 5 V. If input A is now held at 5 V, and input B is again the string of pulses, how does the output change? (In other words, under which condition do the pulses "pass through"? When input A is at 0 V or at +5 V?) Why is this bug called an "AND" gate?

CONSTRUCTION OF A DIGITAL STOPWATCH

25.11. THE DIGITAL STOPWATCH–WHAT'S NEEDED

Suppose you broke the stopwatch you were using for various timing functions during the first semester. You can construct a self-contained digital circuit that duplicates that function, at least for crude measurements. The easiest way to construct a stopwatch is to use several digital integrated circuits along with a few resistors and capacitors.

In order to build a digital stopwatch you will need to use several different IC bugs, including the 4049 inverter and the 4081 AND gate that you investigated in the last several activities. Additional bugs that are needed include the 555 oscillator, a 4026 counter, and an ECG3057 7-segment display. Before starting construction of the stopwatch, let's explore how these other bugs operate.

Fig. 25.15.

25.12. THE 555 OSCILLATOR

In order to build a stopwatch we need a series of pulses that can be passed through an AND gate when its input A is switched "on." You can do this by wiring up an oscillator circuit that spontaneously produces a series of pulses of the kind described above. This circuit employs the popular "555" timer (Sylvania's ECG955), which can generate a series of pulses at a frequency determined by the values of the capacitor and resistors connected to it. In order to investigate the behavior of a 555 timer as an oscillator you will need the following items:

- 1 555 timer (an alternative # is ECG955)
- 1 potentiometer, 10 kΩ
- 1 resistor, 1 kΩ
- 1 electrolytic capacitor, 10 μF
- 1 powered protoboard (with +5 V, +15 V and −15 V) (or a small protoboard and a collection of batteries)
- assorted small lengths of wire (for use with protoboard)
- 1 oscilloscope

Recommended group size:	2	Interactive demo OK?:	N

Wire the circuit as shown in Figure 25.16.

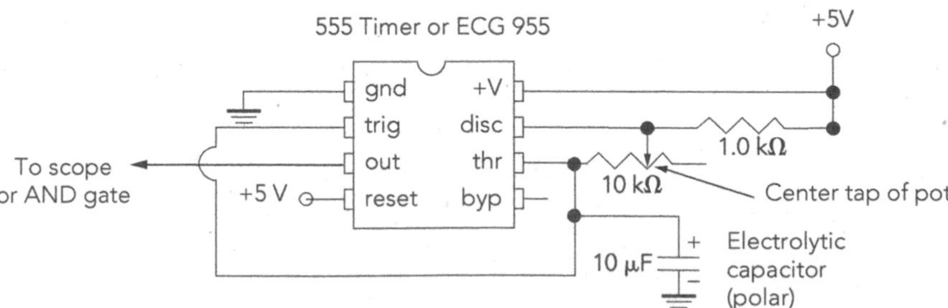

Fig. 25.16. A 555 integrated circuit wired as an oscillator.

25.12.1. Activity: Frequency Output of the 555 Timer

Observe the output signal of the timer wired as an oscillator with the oscilloscope. Sketch and describe what you see. What happens when you change the variable resistor (the 10KV pot)? Turn the 10KV pot to its full resistance. *Calculate the frequency of the pulses.* **Hint:** How long does it take to trace out one full period of a pulse on the scope screen? What is the relationship between frequency and period?

25.13. COUNTING AND DISPLAYING PULSES

If you are to construct your own stopwatch, you need to count pulses coming out of your 555 oscillator and display the number of pulses that are counted. The ECG3057 is a "seven-segment display" that displays a numeral that depends on the potential differences applied to the connecting pins. The "4026" is a counter; when connected properly to the ECG3057, it produces a numeral on the display that increases by one for each electrical pulse that enters the "clock" input of the 4026. After ten pulses, the chip produces another output pulse that can be counted by the next "4026," thus allowing for counts beyond nine. In order to count and display pulses from the 555 oscillator circuit you will need the following additional items:

- 2 seven-segment display elements, 3057
- 2 counters, 4026
- 2 resistors, 1 kΩ
- 1 digital multimeter (to check circuits)

Recommended group size:	2	Interactive demo OK?:	N

In order to test this numeral display circuit, connect the output of the ECG955 oscillator chip directly to the CLK input of the first (right-hand) 4026 counter chip. Then connect the output of the first 4026 counter chip to the

CLK input of the second counter chip. Each of the counter chips must be connected to an ECG3057 display chip. *You can do one of these and your partner the other.*

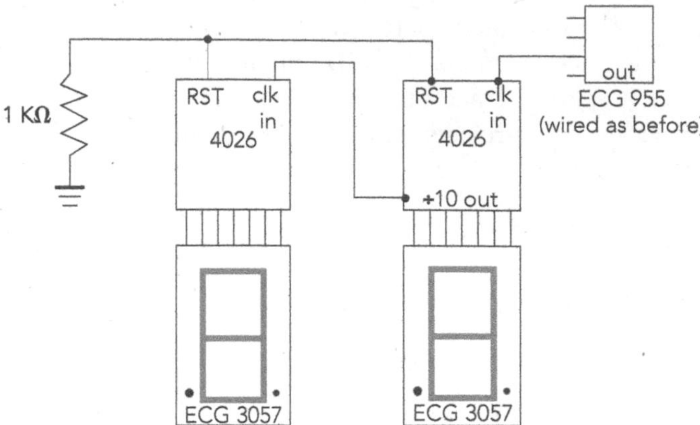

Fig. 25.17. A wiring diagram for seven segment displays.

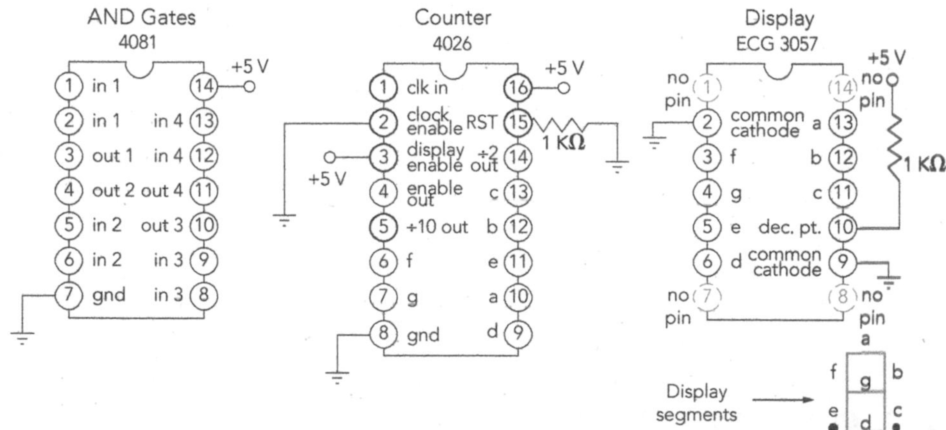

Fig. 25.18. Integrated circuit pinout diagrams.

As the oscillator puts out a string of pulses, the counter should be observed to count rapidly through the numerals 0 to 9, the second numeral display increasing by 1 every time the first one returns to 0. This circuit is a two-digit counter.

25.14. DESIGNING A STOPWATCH

By using the oscillator/counter you just built, along with other circuit elements used in this unit, you can build a stopwatch that will measure elapsed time while a spring-loaded switch is held in the "on" position, indicating seconds and tenths of seconds. For this final activity you will need:

- 1 micro switch
- 1 resistor, 1 kΩ
- 1 capacitor, 0.1μF
- 1 AND gate, 4081
- 1 digital multimeter (to check circuits)

Recommended group size:	2	Interactive demo OK?:	N

Ideally, the stopwatch will reset to zero every time it starts measuring a new time interval. This can be done by switching the RST inputs on the counters to +5 V temporarily and then back to the configuration shown in the circuit diagram above.

Use the circuit diagram shown below to wire your stopwatch. Then check and describe the operation of the circuit. Try measuring some time intervals.

Stopwatch Circuit

In the following circuit, the oscillator is adjusted to produce pulses at a frequency of 10 Hz, so that the counter will count tenths of seconds and, every ten tenths, increase the "seconds" indicator by one. The AND circuit acts as a gate, allowing the 10 Hz pulses through only when the other input is held at 5 V. The capacitor/resistor combination between the switch and the RST (reset) terminal on the counters produces a brief positive pulse when the switch is first pressed; that positive pulse resets the counters to zero before they start counting the 10 Hz pulses.

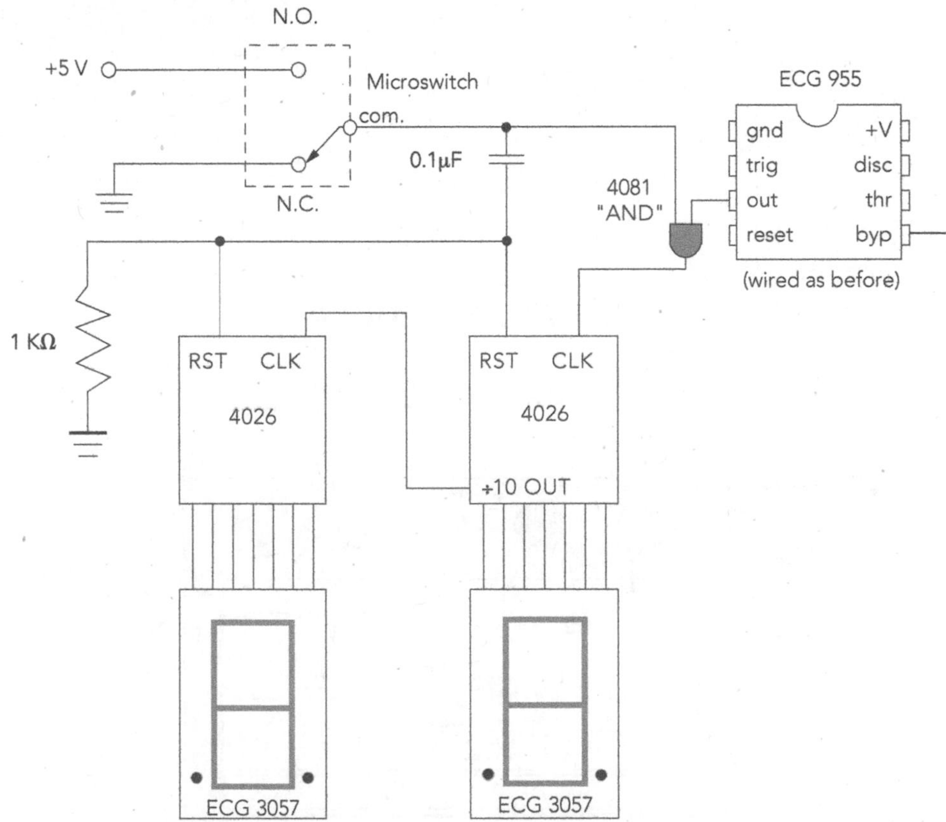

Fig. 25.19. A partial wiring diagram for the digital stopwatch.

25.14.1. Activity: Explaining Your Digital Stopwatch

a. When you and your partner get your digital stopwatch working, show it to your instructor or TA and get a signature witnessing that the stopwatch really works.

I ATTEST THAT THIS STOPWATCH REALLY WORKS!

[Name] ______________________________ [Date]__________

b. Explain as clearly as possible what each part of your circuit does and how each part goes together to make the whole circuit.

UNIT 25 APPENDIX

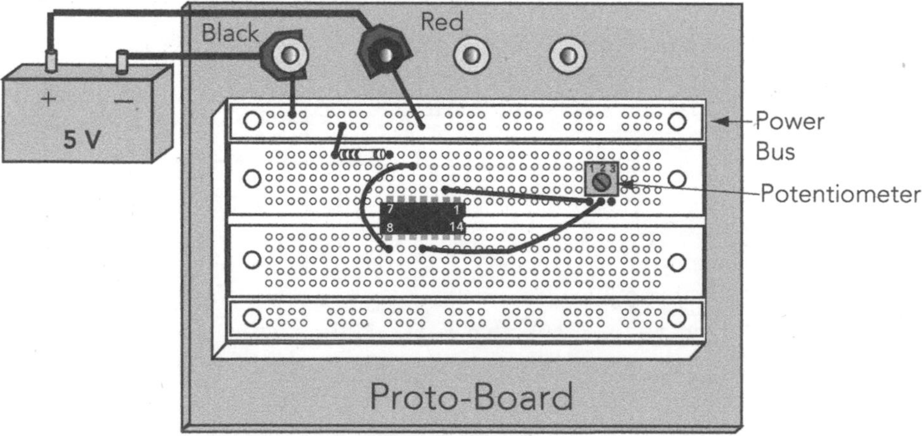

Fig 25.20. Hypothetical sample of how components might be wired using a proto-board. The long strips of dots are usually used for power. The negative or ground terminal of a battery hooked to the outside row of dots and the positive terminal of the battery hooked to the inside row of dots. In this sample, one end of a carbon resistor is hooked to + 5 V (relative to ground) and the other end is hooked to pin 7 of an integrated circuit chip. Meanwhile, pin 2 of the chip is connected to pin 1 of a potentiometer and pin 11 of the chip is connected to the central tap of the potentiometer (pin 2).

Name ______________________ Section ____________ Date ____________

UNIT 26: MAGNETIC FIELDS

Levitation! This is the stuff magicians' dreams are made of. To the physicist this floating magnet is an example of the "magic" of action-at-a-distance resulting from magnetic interaction forces between a small permanent magnet and a low-temperature superconductor carrying a current. What is magnetism? What is the mathematical nature of forces between magnets and moving charges? Why does the small magnet float? When you complete this unit you should be able to answer some of these questions.

UNIT 26: MAGNETIC FIELDS

To you alone . . . who seek knowledge, not from books only, but also from things themselves, do I address these magnetic principles and this new sort of philosophy. If any disagree with my opinion, let them at least take note of the experiments . . . and employ them to better use if they are able.

Gilbert, 1600

OBJECTIVES

1. To learn about the properties of permanent magnets and the forces they exert on each other.

2. To understand how magnetic field is defined in terms of the force experienced by a moving charge.

3. To understand the principle of operation of the galvanometer—an instrument used to measure very small currents.

4. To be able to use a galvanometer to construct an ammeter and a voltmeter by adding appropriate resistors to the circuit.

5. To understand the mathematical basis for predicting that a charged particle moving perpendicular to the magnetic field lines in a uniform magnetic field will travel in a circular orbit.

6. To use an understanding of the forces on an electron moving in a magnetic field to measure the ratio of its charge to its mass, e/m.

26.1. OVERVIEW

Fig. 26.1. Audio cassette.

As a child, you may have played with small magnets and used compasses. Magnets exert forces on each other. Parts of the small magnet that comprises a compass needle are attracted by the earth's magnetism. Magnets are used in electrical devices such as meters, motors, and loudspeakers. Magnetic materials are used in cassette tapes and computer disks. Large electromagnets consisting of current-carrying wires wrapped around pieces of iron are used to pick up whole automobiles in junkyards.

From a theoretical perspective, the fascinating characteristic of magnetism is that it is really an aspect of electricity rather than something separate. In the next two units you will explore the relationship between magnetic forces and electrical phenomena. Permanent magnets can exert forces on current-carrying wires and vice versa. Electrical currents can produce magnetic fields and changing magnetic fields can, in turn, produce electrical fields. In contrast to our earlier study of electrostatics, which focused on the forces between resting charges, the study of magnetism is at heart the study of the forces acting between moving charges.

MAGNETIC FORCES AND FIELDS

26.2. PERMANENT MAGNETS

The attraction of iron to a magnet is so familiar that we seldom realize that most of us know little more than the ancients about how the attraction occurs. Let us begin our exploration of magnetism by playing carefully and critically with some permanent magnets and observing what happens. For the activities involving permanent magnets you will need:

- 2 rod-shaped magnets (with "like" ends marked)
- 1 aluminum conducting rod (the same size and shape as the magnets)
- 5 tiny compasses
- 1 paper clip
- 1 pen
- 1 pencil
- 2 plastic objects
- 2 strings, 10 cm
- 1 Scotch tape, approx. 3″ long
- 1 rod stand
- 2 aluminum rods
- 1 aluminum right angle clamp

Recommended group size:	2	Interactive demo OK?:	N

Interactions Involving Permanent Magnets

Permanent magnets can interact with each other as well as with other objects. Let's explore the forces exerted by one magnet on another and then branch out to make qualitative observations of the forces that a magnet can exert on other objects.

26.2.1. Activity: Permanent Magnets and Forces

a. What do you predict will happen if you bring "unlike" ends together?

b. Do you expect the "like" ends of a magnet will attract or repel each other?

c. Fiddle with the two permanent magnets. Do the like ends attract or repel each other? How do the rules of attraction and repulsion compare to those for electrical charges of like sign? Are the rules the same or different? (One of the ends of a bar or cylindrical magnet is usually called the north pole and the other the south pole. You will explore why shortly.)

d. Each pole represents a different type of magnetic charge. Can you find a magnet with just a north pole or just a south pole? Can you find unlike electrical charges separately? Discuss the differences between electrical and magnetic charges.

e. List at least four objects, one of which is an aluminum rod. *Predict* what will happen if you bring each object near one pole of a magnet and then near the other pole of that magnet.

	Object	Pole 1	Pole 2
1			
2			
3			
4			

f. *Observe* what happens when you bring the various objects close to each of the poles of the magnet and summarize your results in the table below.

	Object	Pole 1	Pole 2
1			
2			
3			
4			

g. How do your predictions and observations compare?

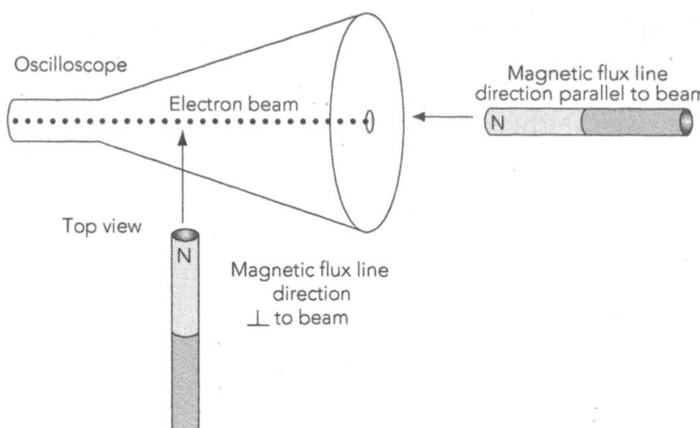

Fig. 26.7. Wire that can carry current in a magnetic field.

26.5.2. Activity: The Magnetic Force Exerted on Moving Charges– An Electron Beam in an Oscilloscope

a. Move the north pole of your magnet parallel and then perpendicular to the electron beam in an oscilloscope. What is the direction of the displacement (and hence the force on the beam) in each case? Sketch vectors showing the direction of the magnetic field, the direction of motion of the original electron beam before it was deflected, and the direction of the resultant force on the beam.

b. You should have found that when the bar magnet is perpendicular to the beam of moving charge, the magnetic force is perpendicular to both the direction of the magnetic flux line and the moving charge. What type of vector product can give a vector that is perpendicular to two other vectors? The dot product or the cross product?

c. Suppose we define magnetic field as a vector quantity $\vec{B}$ that points along the axis of the magnet away from the north pole and toward the south pole. Show that the vector cross product $\vec{F}^{mag} = q\vec{v} \times \vec{B}$ properly describes your observations (at least qualitatively) in terms of the relative directions of the three vectors. **Hint:** Don't forget that q is negative in the case of an electron beam.

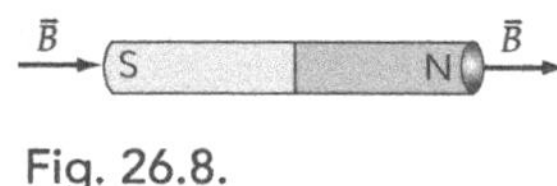

Fig. 26.8.

The force that results from the movement of charge in a magnetic field is known as the Lorentz force. This leads to a rather backwards mathematical definition of magnetic field, $\vec{B}$. The magnetic field, $\vec{B}$, is defined as that vector that, when crossed into the product of charge and its velocity, leads to a force, $\vec{F}^{mag}$, given by the cross product $\vec{F}^{mag} = q\vec{v} \times \vec{B}$.

The standard unit for $\vec{B}$ that follows from this cross product in newtons per coulomb-meter per second. *This unit of magnetic field is known as the tesla, or T for short.*

26.6. REVIEW OF THE VECTOR CROSS PRODUCT

Once again we have this weird mathematical entity called the *vector cross product* that you encountered in Unit 13. You have probably utterly forgotten what the symbols mean. Let's review the cross product. The force on a moving charge q in a magnetic field can be described mathematically as the "vector cross product" that has:

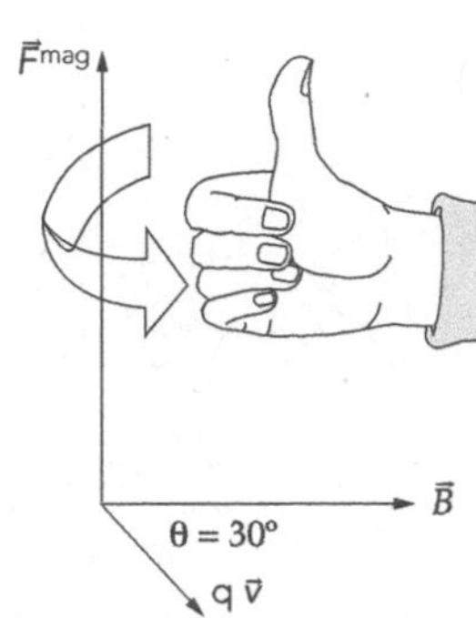

Fig. 26.9. Use of the Right-Hand Rule to find the direction of a magnetic force exerted on a moving charge. The $\vec{v}$ and $\vec{B}$ vectors are 30° apart and in a plane ⊥ to the paper. The $\vec{F}$ vector is ⊥ to both $\vec{v}$ and $\vec{B}$.

1. Magnitude: The magnitude of the cross product is given by $|q|vB \sin\theta$ where θ is the smallest angle between the velocity vector, $\vec{v}$, and the magnetic field vector, $\vec{B}$. (As usual, v and B represent vector magnitudes.)
2. Direction: The cross product, $\vec{F}^{mag}$, is a vector that lies in a direction ⊥ to both $\vec{v}$ and $\vec{B}$ and is "up" (for positive q) when the fingers of the right hand curl from $\vec{v}$ to $\vec{B}$ with the thumb up. The cross product is "down" (for positive q) when the fingers of the right hand curl from $\vec{v}$ to $\vec{B}$ with the thumb down. If q is negative, then the direction is reversed in each case. The properties of the cross product are pictured in Figure 26.9.

Since, the spatial relationships between $\vec{v}$, $\vec{B}$, and $\vec{F}^{mag}$ are difficult to visualize. In the next activity you can practice using the cross product.

26.6.1. Activity: Using the Lorentz Force in Calculations

a. Consider a proton traveling at $\theta = 30$ degrees with respect to a magnetic field of strength 2.6×10^{-3} T as shown in Figure 26.9. It has a speed of 3.0×10^{6} m/s. What is the magnitude and direction of the force exerted on the proton by the magnetic field?

b. If the particle is an electron instead, what is the magnitude and direction of the force exerted on the particle by the magnetic field?

MAGNETIC FORCES AND ELECTRIC CURRENTS

26.7. MAGNETIC FORCE ON A CURRENT-CARRYING WIRE

You demonstrated earlier that permanent magnets do not exert forces on the static charges on a piece of charged Scotch tape. On the other hand, magnetic fields from a permanent magnet can bend a beam of electrons. What do you think will happen to a non-magnetic wire that is placed near a magnet? How about when the wire carries a current? To examine the effects of a magnet on a wire you will need:

- 1 lantern battery, 6 V
- 2 alligator clip leads, 20 cm
- 2 alligator clip leads, 10 cm
- 1 SPST switch
- 1 lab magnet, 7.5 KGauss or more

Recommended group size:	3	Interactive demo OK?:	Y

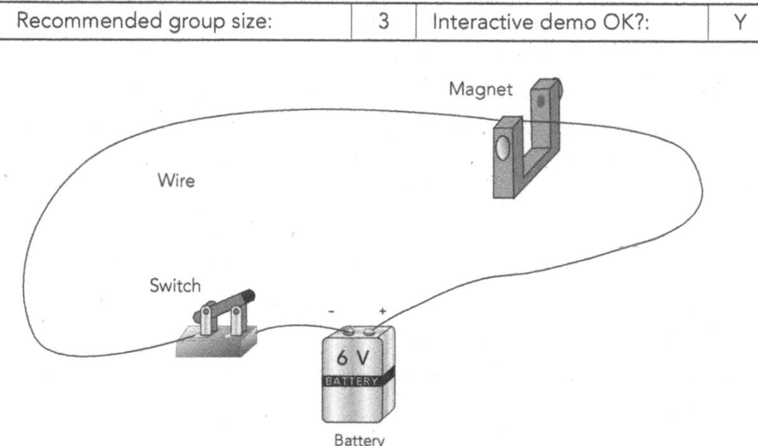

Fig. 26.10. Wire that carries current in a magnetic field when the switch is closed.

26.7.1. Activity: Predicted and Observed Forces on a Wire

a. Use the Lorentz force equation to predict the nature of the force on a non-magnetic wire when it is placed between the poles of a strong permanent magnet so that part of the wire is in a fairly strong magnetic field. There is no need to be quantitative; a qualitative explanation is fine.

1. When the wire carries no current:

2. When the wire carries a current:

b. Set up the circuit shown in Figure 26.10 and hold the wire between the poles of the magnet while your partner opens and closes the switch. **Warnings:** (1) Keep the alligator clip leads away from the magnet's poles as they have iron in them. (2) Do not leave the switch on for long or the batteries will go dead quickly. Describe your actual observations. If there is a force on the wire, note the *direction* of the force relative to the direction of the B-field and the current. You may want to change the direction of the current by reversing the batteries or the direction of the wire.

1. When the wire carries no current:

2. When the wire carries a current:

c. Are your observations consistent with the Lorentz force law? Why or why not?

26.8. MAGNETIC FORCE ON A CURRENT LOOP

The galvanometer, which is made up of a current loop and a permanent magnet, lies at the heart of many electrical measuring devices. In order to understand its operation, let's consider what happens when a rectangular loop carries current in the presence of a magnetic field.

Assume there is a current i flowing around the rectangular loop shown in Figure 26.11. Since current consists, by definition, of moving electrical charges, any magnetic field at right angles to the current should exert a Lorentz force on the wire. Assume that the current loop is in the plane of the paper. A magnetic field, parallel to the plane of the paper, is present. The loop can pivot about the line CD.

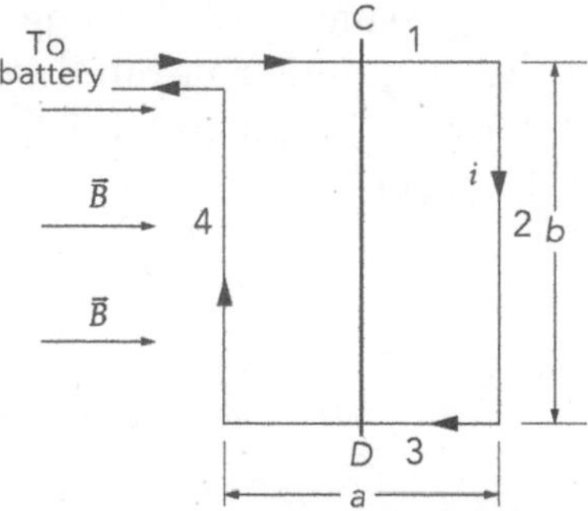

Fig. 26.11. One loop of a multiloop wire of dimensions a, b in a magnetic field of magnitude B.

You can answer the questions below with a combination of direct observations and mathematical reasoning. For your observations you will need:

- 1 rectangular wire loop, R approx. 2 Ω (with one turn or multiple turns)
- 1 battery, 4.5 V
- 3 alligator clip leads, approx. 20 cm
- 1 U-shaped magnet, 2.5″ gap
- 1 small compass
- 1 ammeter
- 1 SPST switch

Recommended group size:	2	Interactive demo OK?:	N

For the required mathematical reasoning assume that a positive current I is made up of a series of positive charges, q, each moving with an average speed v in the direction of i.

26.8.1. Activity: Predicted Forces on a Current Loop

a. Use the Lorentz force equation to show mathematically that an electron with a negative charge of $q_e = -e$ (or -1.60×10^{-19} C) moving at a velocity $\vec{v}$ will experience the same force in a magnetic field $\vec{B}$ the same amount of positive charge $-q_e = +e$ (or $+1.60 \times 10^{-19}$ C) would experience moving at a velocity $-\vec{v}$. Be sure to use the vector signs in writing your symbols.

b. Use the Lorentz force law and the right-hand rule to determine the theoretically predicted direction of the magnetic force on each segment of the wire (1, 2, 3, and 4). For simplicity, assume that the current i consists of *positive charges*, q, moving at an average speed, v, in the *same direction* as i. Assume that the loop lies in the plane of the paper as shown in the diagram. You can describe the directions of these forces in such terms as "right-to-left," "left-to-right," "into the paper," and "out of the paper."

c. What motion of the loop do you predict will result from these forces?

d. Suppose the plane of the wire loop is rotated 90° about the axis through points C and D so it is *perpendicular* to the plane of the paper with side 2 above the paper and side 4 below. Assume $\vec{B}$ is still in the plane of the paper. (What is the predicted direction of the forces on each segment of the wire?)

e. What should happen to the force on each segment of wire when the current in the loop increases? Should it increase, decrease, or stay the same? Why?

In order to observe the forces on a current loop we will need to put the loop into a magnetic field that has a definite direction and pass an electric current through it. We will use the space between the poles of a U-shaped magnet for the magnetic field. The forces will be larger if you use a wire with multiple loops on it. Why?

26.8.2. Activity: The B-field Inside a U-shaped Magnet

a. If a U-shaped magnet is just a bar magnet bent into the shape of a U, what do you predict is the direction of the magnetic field between its poles? Sketch the lines. **Hint:** Take a look at the field lines from your straight magnet.

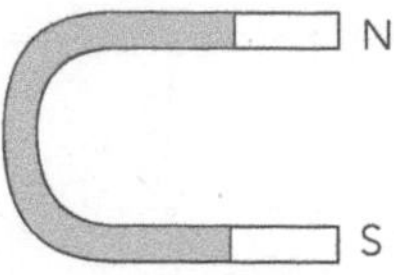

b. Place a small compass between the poles of a U-shaped magnet and sketch the actual field line directions in the diagram below.

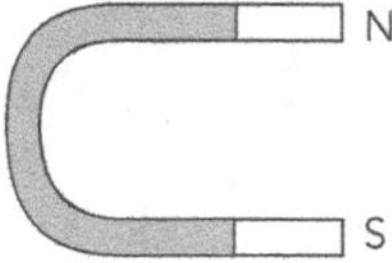

Next let's check out your predictions for the force on a loop placed at various angles between the poles of a U-shaped magnet. For the required observations you should wire a single 6 V battery, rectangular loop, and switch in series.

26.8.3. Activity: Observed Forces on a Current Loop

a. Place the loop of wire so its plane is parallel to the magnetic field between the poles of the magnet. Pass current through the loop. What happens? How does this compare with your prediction?

b. What happens when you add another battery to the circuit so the current increases? What happens if you reverse the battery's polarity so the current travels through the loop in the opposite direction?

c. Explain how you might use this setup to measure current.

d. Place the loop of wire so its plane is perpendicular to the magnetic field between the poles of the magnet. Pass current through the loop. What happens? How does this compare with your prediction?

26.9. THE GALVANOMETER

A measuring instrument called the galvanometer consists of a series of wire loops placed in a magnetic field while current passes through them. Although digital electronic meters work differently, the galvanometer lies at the heart of many old fashioned high precision voltmeters and ammeters. A diagram of a typical galvanometer is shown in Figure 26.12.

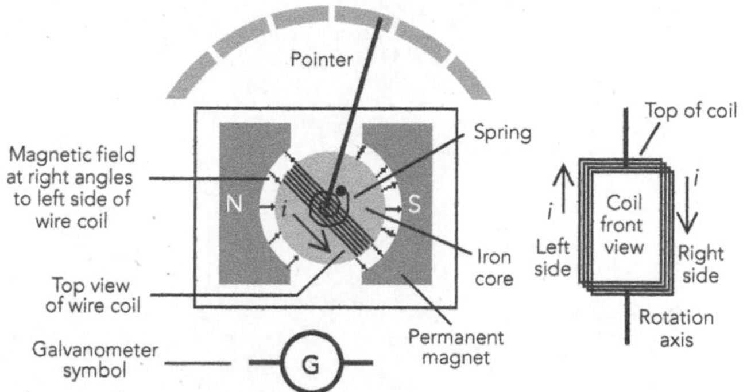

Fig. 26.12. The guts of a typical galvanometer.

The ability of the new digital multimeters to make accurate current measurements for small currents is poor. Let's construct a more accurate ammeter using a galvanometer. For this activity you will need the following:

- 1 galvanometer
- 4 alligator clip leads, approx. 20 cm
- 1 battery, 4.5 V
- 1 ammeter, 3 mA
- 1 SPST switch
- 1 set of carbon resistors (with a full range of standard values)
- 1 multimeter

Recommended group size:	4	Interactive demo OK?:	N

Your galvanometer is just an ammeter capable of measuring very small currents. The internal resistance of a galvanometer can be represented by the symbol R_g, where the value of R_g is usually listed on the meter. Most galvanometers also list the full-scale-deflection current. Some galvanometers list the amount of current that will cause one division of deflection; the full-scale-deflection current, i_{fs}, can easily be determined from this information.

26.9.1. Activity: Using a Galvanometer to Make an Ammeter

a. Let's summarize the characteristics of your galvanometer in preparation for constructing an ammeter that can measure a current that is up to 5 times the full scale deflection current of your galvanometer.

Internal resistance: $R_g =$ ______ Ω

Full-scale-deflection current: $i_{fs} =$ ______ × ______ = ______ A

(Number of divisions × Current/division)

5 times the full-scale-deflection current: $i^{max} =$ ______ A

b. Design an ammeter that reads a maximum current, i^{max}, which is 5 times the full-deflection-scale current, i_{fs}. To accomplish this we must put a small resistor in parallel with the galvanometer to "shunt" off most of the current. Examine the circuit diagram that follows, and then show mathematically that the value of the shunt resistance, Rs, must be given by the equation

$$R_s = \frac{i_{fs} R_g}{(i^{max} - i_{fs})}$$

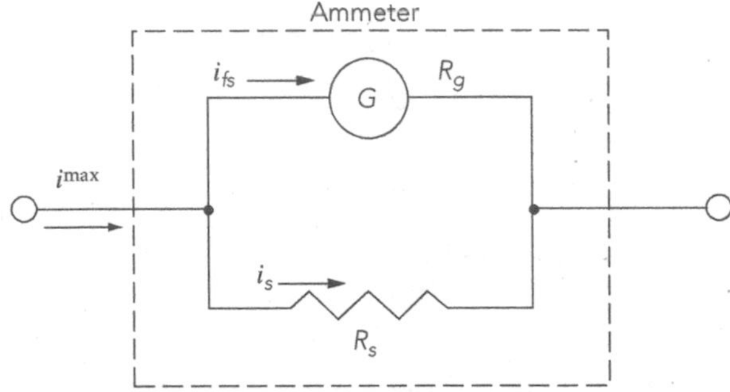

Fig. 26.13. Wiring an ammeter unit using a galvanometer and shunt resistor.

c. Use the equation you just derived to calculate the value of the shunt resistance that you need to use to have i^{max} be 5 times the full deflection scale current, i_{fs}.

Shunt Resistance: $R_s =$ ______ Ω

d. You can test your new galvanometer-based ammeter by wiring it in series with a resistor and battery as shown in the circuit in Figure 26.14.

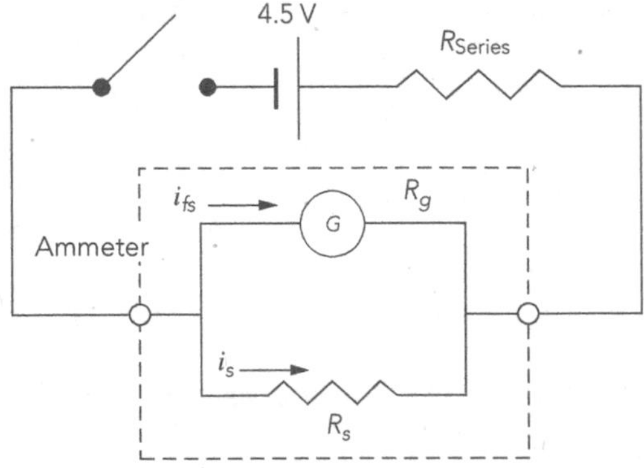

Fig. 26.14. Ammeter in series with a resistor and battery.

Before you actually wire up the circuit to test your ammeter, choose a shunt resistor with a resistance as close as possible to the one you calculated in part d. Measure its value with a multimeter and also measure the actual potential difference across your battery.

Actual value of the shunt resistance: $R_s =$ ______ Ω

Actual battery potential difference: $\varepsilon =$ ______ V

e. Next, calculate the value of i^{max} corresponding to the value of your actual shunt resistance, R_s. Then calculate the value of the resistor you will need in series with your battery to limit its current to i^{max} or less.

Maximum current that can be measured: $i^{max} =$ ______ A

Series resistance: $R_{series} =$ ______ Ω

f. Test your circuit using the resistor, R_{series}, in series with the 4.5 V battery to reduce the circuit current to i^{max} or less. Use another ammeter to verify your meter reading. Did your ammeter work? Explain what you saw on the galvanometer.

26.10. MAGNETIC FORCES ON A CURRENT-CARRYING CONDUCTOR

Since current represents a collection of many charged particles in motion, a current-carrying wire experiences a force in the presence of a magnetic field unless the field is parallel to the wire. In order to use the Lorentz force law to calculate the force on a conductor of length L carrying a current, you need to relate the average or drift velocity, $<\vec{v}>$, of the charge flowing through the wire to the current in the wire and the length of the wire. Suppose a charge q travels through a straight wire of length L in an average time interval of Δt.

26.10.1. Activity: Relating the Lorentz Force to Current

a. What is the equation for the average velocity $<\vec{v}>$ of the charges in terms of the length vector, $\vec{L}$, which has a magnitude equal to the length of a straight wire segment and the same direction as the motion of positive current?

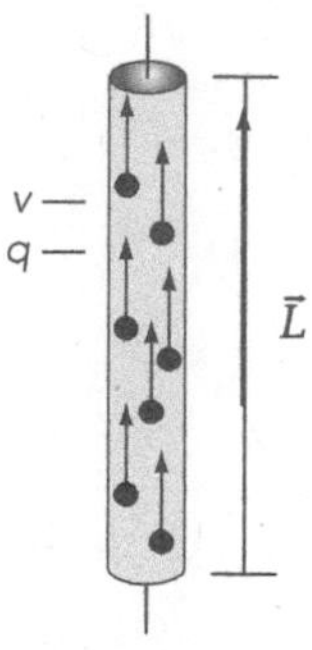

Fig. 26.15.

b. What is the equation for the wire current, i, in the wire in terms of q and Δt?

c. Show that $q<\vec{v}> = i\vec{L}$ and that the Lorentz force on a straight current-carrying wire of length L is given by the expression $\vec{F}^{mag} = q\vec{v} \times \vec{B} = i\vec{L} \times \vec{B}$.

d. Suppose a wire is not straight. Use the expression in part c. above to express the magnetic force, $d\vec{F}^{mag}$, on a small, almost straight, segment of length $d\vec{L}$ in the presence of a magnetic field $\vec{B}$.

FORCES AND FIELDS CAUSED BY CURRENTS

26.11. MAGNETIC FORCES ON A SEMICIRCULAR WIRE

Consider the length of current-carrying wire with a semicircular kink in it shown in Figure 26.16. Predicting and calculating the force on the wire when placed in a magnetic field is good practice in the use of physical reasoning about the action of the Lorentz force and the use of the spreadsheet to do numerical calculations. You'll need to use the equation you derived in Activity 26.9.1d for your calculations.

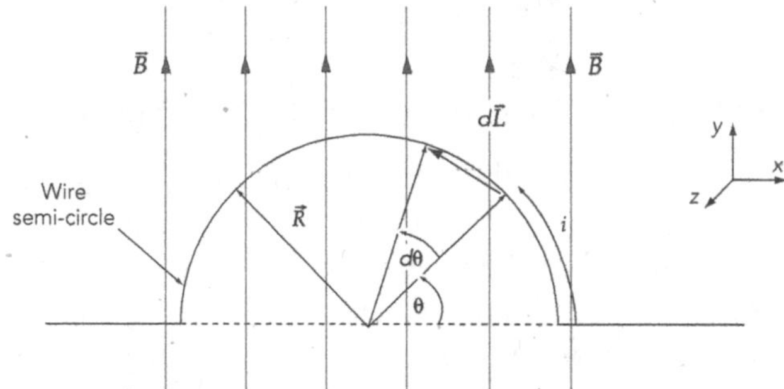

Fig. 26.16. A semicircular segment of current-carrying wire in a uniform magnetic field. The size of the infinitesimal increase in angle $d\theta$ is exaggerated.

26.11.1. Activity: Magnetic Force on a Kinky Wire

a. Assume that the magnetic field is positive in the $+y$ direction and that positive current is traveling counterclockwise. Use qualitative reasoning to determine the direction of the net force on the wire. Explain your reasoning.

b. Use the equation you derived in Activity 26.10.1d. Break the wire into at least 15 segments and perform a numerical calculation of the net force on the wire using the spreadsheet. (Your spreadsheet should look like the following sample.) Assume that $B = 10.0$ teslas over the region, $i = 0.5$ A in a counterclockwise direction. Suppose the radius of the semicircular kink is given by $R = 7.0$ cm. Plot the force on each segment of wire as a function of the segment number. Why is the force much smaller for some segments than for other segments? Affix your printouts in the following space.

$i =$	...
$B =$	...
$\Delta l =$	...

Seg #	ø	sin ø	ΔF
1	6	...	...
2	18	...	...
3	30	...	...
.	.	.	.
.	.	.	.
.	.	.	.
15	174	...	...
		force =	...

26.12. OVERVIEW: CHARGED PARTICLES IN A MAGNETIC FIELD

A charged particle moving in a magnetic field experiences a Lorentz force perpendicular to its velocity. In the next activity you are going to study a famous piece of apparatus used to measure the charge on the electron known as the "e/m apparatus"—pronounced by physicists as "eee-over-mmm." The reason for the name is that it can be used to determine the ratio of e/m, but not the value of e or m individually. This may seem like small potatoes to you, but back when physicists were trying to determine these actual quantities that we all take for granted today, it was difficult to come up with exact values for either e or m for such a tiny fundamental particle as the electron!

In an e/m apparatus, a beam of electrons is accelerated by a large potential difference to a fairly high velocity. The beam is then passed into a magnetic field that is perpendicular to the direction of motion so that it experiences a Lorentz force given by the now familiar ultra-perpendicular equation

$$\vec{F}^{mag} = -e\vec{v} \times \vec{B}$$

where e is the symbol used to represent a positive quantity signifying the amount of charge of the electron (so that $q_e = -e$). The way in which the electron beam bends in a known magnetic field allows us to determine the ratio of electron charge to the electron mass, e/m, experimentally.

This exploration of the motion of electrons in a uniform magnetic field will involve several activities: (1) first, you'll predict what kind of path an electron will follow if it is shot into a uniform magnetic field; (2) next, you'll swing a mass in a circle on a string to get a feel for how the force you have to apply to the mass to keep it moving in a circle influences its radius and speed; (3) then you'll use the results of the first two activities to figure out how you can measure e/m if an electron beam of known velocity is bent by a uniform magnetic field into a circle of measurable radius; (4) you'll determine that current-carrying coils can produce a magnetic field that is fairly uniform; and, finally, (5) you'll use a classic e/m tube placed in a magnetic field to measure the ratio of charge to mass, e/m, for the electron.

Path of an Electron in a Magnetic Field

Consider an electron that is shot with velocity $\vec{v}$ from left to right in the presence of a uniform magnetic field $\vec{B}$ that is into the paper. This is indicated by the X's in Figure 26.17.

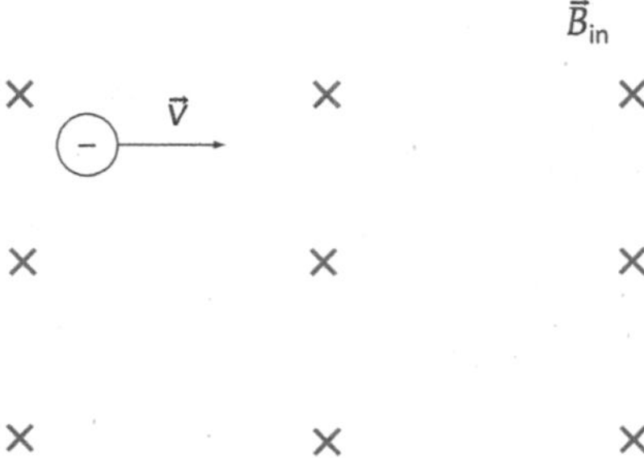

Fig. 26.17. An electron with negative charge given by $q_e = -e$ traveling in a magnetic field that is perpendicular to its direction of motion.

26.12.1. Activity: Magnetic Force on an Electron

a. Using a qualitative argument based on the Lorentz force law, what is the direction of the force on the electron? You can use the terms up, down, L-R, R-L, or in, out to describe the direction of the force. Sketch a force vector on the charge on Figure 26.17. **Hint:** Don't forget the electron has a negative charge.

b. In the next moment after it is launched, will the electron still be traveling in the same straight line? Why or why not? Sketch where it might be in the next moment on Figure 26.17.

c. If the *Lorentz force* is perpendicular to the direction of motion of the electron in the first moment, is it still perpendicular in the second moment? In the third moment? Why or why not?

d. If the force is always perpendicular to the direction of motion, is any work done on the particle as it moves in a curved path? Recall that the formal definition of work for a small displacement $d\vec{s}$ is given by the equation $dW = \vec{F} \cdot d\vec{s}$.

e. If no work is done on the electron as it moves, does its speed (that is, the magnitude of its velocity) change or remain the same?

f. The displacement of an electron bending in a magnetic field is shown in Figure 26.18 for the first two moments. Complete the diagram and thus show the shape of the path of the electron in the magnetic field.

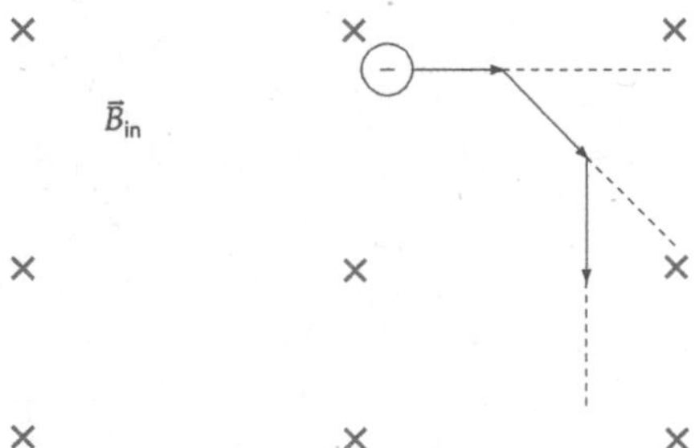

Fig. 26.18.

g. Suppose, like Mother Nature, you broke the path above up into a huge number of tiny steps. What would the shape of the path be? How might it change if you increase the magnitude of the magnetic field?

26.13. THE FORCE NEEDED TO MAINTAIN CIRCULAR MOTION

In order to relate the radius of a circular path of an electron beam traveling in a magnetic field to the charge to mass ratio, e/m, of the individual electrons, let's explore the forces needed to keep a mass moving in a circle. For this activity you'll need the following items:

- 1 string, 1.5 m
- 1 ball (attached to the string)

Recommended group size:	2	Interactive demo OK?:	Y

Fig. 26.19. Olympic silver medalist Paul Wylie is travelling in a circular path. This circular motion is not possible unless the ice exerts a force on him directed toward the center of curvature. His lean toward the center of curvature allows the horizontal component of force from the ice to transmit a centripetal force to his body.

You can go outdoors and swing the ball overhead in a circle at different radii and speeds. As you swing the ball you should compare the force you need to keep the ball moving in a circle under different circumstances.

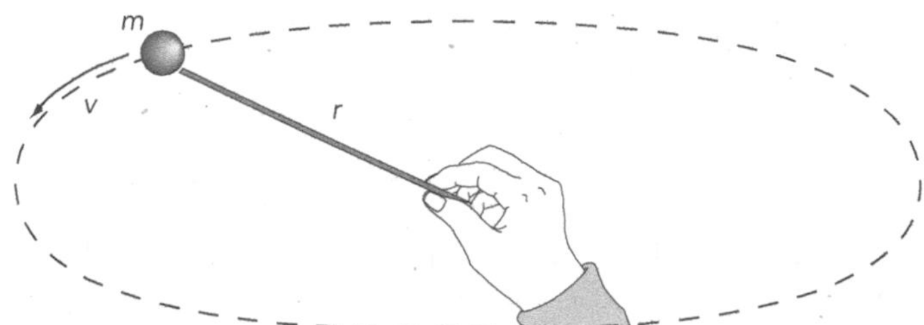

Fig. 26.20. Swinging a mass in a circle.

26.13.1. Activity: Circular Motion–F, r and v

a. Hold a string tightly as you swing a ball around in a circle at several different speeds. Does the force you need to exert on the string change as you increase the speed of the ball? What do you predict the mathematical relationship between speed and force will be if you were to take careful measurements?

b. Hold the speed of the rotating ball constant and swing the ball around in a circle at two different radii. First, swing the ball at a radius r with a period T. Then swing it at a radius $2r$ and keep the speed constant by allowing the period T to double to a time $2T$. Does the force you need to exert on the string with your hand to keep the ball from flying off stay the same or increase as you increase the radius of the ball? What do you predict the mathematical relationship between radius and force will be if you were to take careful measurements? Explain.

c. Suppose you were to tie a bowling ball to the end of your string. Would you expect to have to exert more force or less force to keep the heavier ball in its circular orbit? What do you predict the mathematical relationship between mass and force will be if you were to take careful measurements?

26.14. DERIVING AN EQUATION FOR *e/m*

OK, now let's put all the pieces together. Assume that we use a potential difference of ΔV to accelerate a beam of electrons to a speed of v. Assume that after acceleration the electron beam is moving perpendicular to a uniform magnetic field B and that we then observe that the electrons move in a circle of radius r. What theoretical equation would you use to calculate a value of e/m from your measurements of B, r, and the accelerating potential difference ΔV?

In Unit 7 you verified that the magnitude of the *centripetal force* needed to keep a particle of mass m moving in a circle of radius r depends on the square of the speed and is given by Equation 7.3

$$F = \frac{mv^2}{r} \qquad \text{(centripetal force magnitude)}$$

26.14.1. Activity: Derivation of the *e/m* Equation

a. Since the Lorentz force always acts inward toward the center of the circular path taken by an electron in a magnetic field, find the equation of e/m in terms of v, r, and B by assuming that the centripetal force is provided by the magnetic Lorentz force. Use the Lorentz force law for $\theta = 90°$ along with the centripetal force equation to determine how e/m depends on B, r, and v.

b. What is the equation for the kinetic energy of a mass m moving at a speed v?

c. The electrons in an e/m tube are accelerated to a speed v by falling through a potential difference of ΔV before being shot into the magnetic field. What is the equation for the potential energy lost by a charge e falling through a potential difference ΔV?

d. Set the potential energy lost by the electron equal to the kinetic energy gained by it as a result of its acceleration. Then use the equation you derived in part a. to show that e/m is give by the expression

$$\frac{e}{m} = \frac{2\Delta V}{r^2 B^2}$$

26.15. FINDING *e/m* EXPERIMENTALLY

Exploring the Properties of the Magnetic Field Produced in an e/m Apparatus

The only remaining task is to figure out how to determine the magnitude of the magnetic field that the beam of electrons moves through. In the typical apparatus used to measure e/m, an approximately uniform magnetic field is produced by two large current-carrying coils called Helmholtz coils. (We'll be studying the ability of currents to produce magnetic fields in the next unit.) For now, you should just accept on faith that the magnetic field between the Helmholtz coils can be calculated from well-accepted equations if the current in the coils is known.

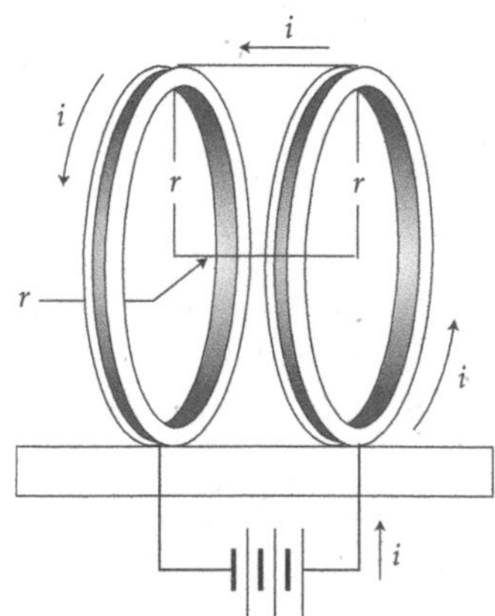

Fig. 26.21. Helmholtz coils to produce a magnetic field.

You can use a small compass to verify that, inside the current-carrying coils, the direction of the magnetic field is perpendicular to the plane of the electron beam. You can also see that the field is fairly strong between the coils and imagine that it is plausible that the field is relatively uniform.

In the next unit you will learn more about the relationship between current in wires and magnetic fields. For a pair of Helmholtz coils of radius R and spacing R as shown in Figure 26.21, it can be shown that the magnitude of the magnetic field, B, in the region between the two coils is given by

$$B = \frac{8\mu_0 Ni}{\sqrt{125}\,R} \text{ teslas}$$

where $\mu_0 = 4\pi \times 10^{-7}$ [N/A^2] (the magnetic constant in air or a vacuum)
N = the number of turns in the coil
R = the radius of each coil and the spacing between them in meters
i = the current through the coils in amps

Measuring *e*/*m*

In order to find e/m experimentally, you will need:

- 1 e/m apparatus with Helmholtz coils
- 1 small compass

Fig. 26.22.

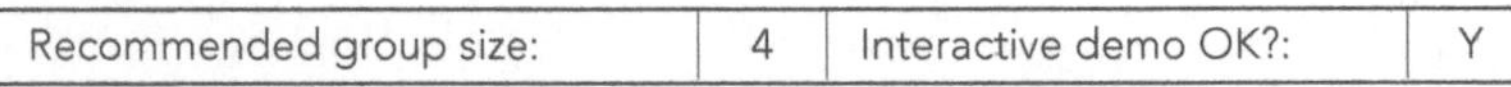

Recommended group size:	4	Interactive demo OK?:	Y

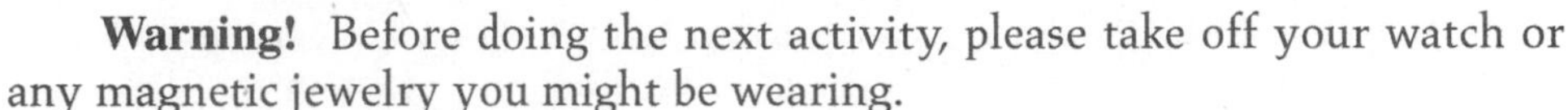

Warning! Before doing the next activity, please take off your watch or any magnetic jewelry you might be wearing.

Your instructor will provide you with information about the recommended current through the Helmholtz coils, the resulting value for the magnetic field between the coils, the dimensions of the tube, and the accelerating voltage for the electron beam.

Measurement	Symbol	Value	Units
Electron beam accelerating voltage	V		volts
Radius of beam path (half the filament-to-pin distance)	r		meters
Helmholtz coil current	i		amperes
Number of turns of coil wire[1, 2]	N		—
Radius of Helmholtz Coil[1, 2]	R		meters

1. For a standard Welch e/m apparatus $N = 72$, $R = .33$ m and tube dimensions are scribed on the brass plate at the bottom of the apparatus as follows: Typical filament to pin distances (that is, $2r$) are: pin #1 (6.48 cm); pin #2 (7.75 cm); pin #3 (9.02 cm); pin #4 (10.3 cm); pin #5 (11.5 cm).

2. For the Pasco SE-9625 Apparatus $N = 130$ and $R = 0.15$ m.

26.15.1. Activity: The Magnetic Field Inside the Coils

a. Turn on the recommended current through the coils and place the compass between the coils. Are the magnetic field lines parallel or perpendicular to the plane of the coils? What is the evidence for your answer?

b. Does the field inside the coils seem fairly uniform? Cite evidence for your answer.

c. Refer to the equation above relating the magnetic field inside the coils to N, i, and R. Use the value of current, i, and the value for the number of turns on each coil to compute the value of the magnetic field, B, in teslas.

Calculating *e/m* from Measurements

After all of this preparation you are finally ready to calculate a value of e/m using measurements.

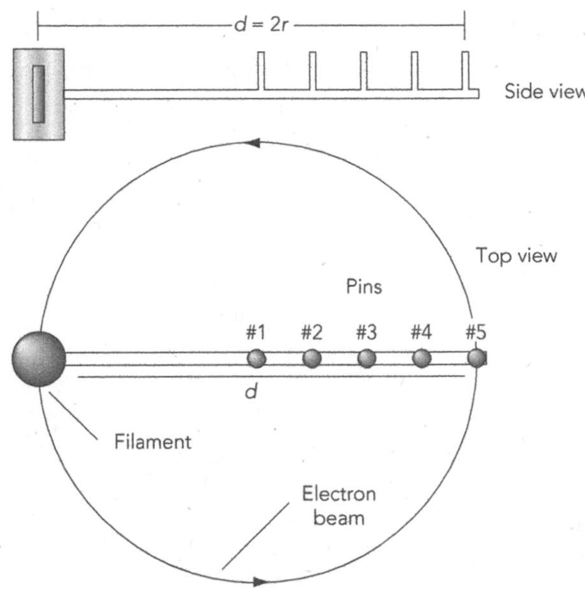

Fig. 26.23. Views of the inside of an *e/m* tube.

26.15.2. Activity: Determining *e/m* Experimentally

a. Use the values for the magnetic field, B, the radius of the electron beam, r, and the recorded value of the accelerating voltage, V, necessary to keep the electron beam moving in a circle (of a diameter equal to the distance between the filament opening where the beam emerges and the pin that the outside of the beam hits as shown in Figure 26.23 above) to calculate the value of e/m. Be sure to include units!

b. Look up the accepted values of e and of m in your textbook and calculate an accepted value of e/m. Note that the accepted values of e and m are also measured values. However, they represent measurements using the best known techniques and apparatus.

c. What is the percent discrepancy between your measured value of e/m and the accepted value of e/m?

Name ______________________ Section ____________ Date ____________

UNIT 27: ELECTRICITY AND MAGNETISM

Small electronic devices like cassette recorders often require between 6 and 9 volts to operate. What's in those heavy cubes we plug into the wall to power electronic devices? It is a transformer, which is used to reduce an input voltage. When a cube like the one in the photograph is plugged into the wall, the 110-volt source causes an alternating current to flow in a wire coil. This current, which alternates direction 60 times per second, produces a magnetic field in an iron core. This changing magnetic field then induces an alternating voltage in another wire coil but this voltage can be lower. In this unit you will learn about the processes by which currents produce magnetic fields and changing magnetic fields in turn induce currents. The design of a power cube is based on these processes.

UNIT 27: ELECTRICITY AND MAGNETISM

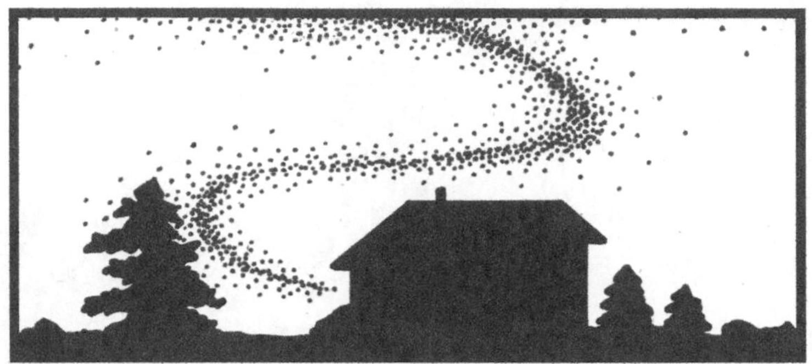

Occasionally during these years . . . [Michael Faraday] thought of electrical problems. One of special interest was the question: Since magnetism can be produced from electricity, can electricity be produced from magnetism? Everything in nature is nicely balanced and symmetrical; in the words of Newton, there is action and there is reaction. Force will give motion; motion will give force. Heat will cause pressure; pressure will cause heat. Chemical action will produce electricity: electricity will produce chemical action. Then, since electricity will develop magnetism, will not magnetism develop electricity? H. H. Skilling

OBJECTIVES

A. Magnetism from Electricity

1. To learn by direct observation about the direction of magnetic field lines produced by a current in a straight wire.

2. To learn about the potential health effects of magnetic fields induced by electrical currents in power lines and home appliances.

3. To understand how to use Ampère's law to calculate the magnetic field around a closed loop in the presence of electrical currents.

B. Electricity from Magnetism

1. To observe that an electric field can be produced by a changing magnetic field by a process known as induction.

2. To explore the mathematical properties of induction as expressed in Faraday's law and to verify Faraday's law experimentally.

27.1. OVERVIEW

In the last unit you observed that permanent magnets can exert forces on both freely moving charges and electrical currents in conductors. We have postulated the existence of a mathematical entity called the magnetic field in order to introduce the Lorentz force law as a way of mathematically describing the nature of the force that a permanent magnet can exert on moving electrical charges. Newton's third law states that whenever one object exerts a force on another object, the latter object exerts an equal and opposite force on the former. Thus, if a magnet exerts a force on a current-carrying wire, mustn't the wire exert an equal and opposite force back on the magnet? It seems plausible that the mysterious symmetry demanded by Newton's third law would lead us to hypothesize that if moving charges feel forces as they pass through magnetic fields, they should be capable of exerting forces on the sources of these magnetic fields. It is not unreasonable to speculate that currents and moving charges exert these forces by *producing magnetic fields themselves.* One of the agendas for this unit is to investigate the possibility that an electrical current can produce a magnetic field.

This line of argument, based on Newton's third law and its symmetry, can lead us into even deeper speculation. If charges have electric fields associated with them, then moving charges can be represented mathematically by changing electric fields. Thus, using the concept of "field" to describe forces that act at a distance, we can say that changing electric fields are the cause of magnetic fields. This leads inevitability to the question: if this is so, then, by symmetry, *can changing magnetic fields cause electric fields?*

This unit deals with two questions. (1) Does a long straight current-carrying wire produce a magnetic field? If so, what quantitative mathematical relationships can be used to describe the nature of such a field? This will lead us to present Ampère's law, which describes the magnetic field around a closed loop as a function of the current enclosed by the loop. (2) How does Faraday's law describe the relationship between changing magnetic fields, the electric fields, and currents they produce?

Faraday's law lies at the absolute heart of the study of electricity and magnetism. It is one of the most profound laws in classical physics. By combining Ampère's law with Faraday's law, we can describe mathematically how electricity produces magnetism and how magnetism produces electricity. Thus, two seemingly different phenomena, electricity and magnetism, can be treated as aspects of the same phenomenon. At the end of this unit, we will peek briefly at the reformulation of some of the laws of electricity and magnetism that we have already learned into a famous set of four equations known as Maxwell's equations.

MAGNETISM FROM ELECTRICITY

27.2. THE MAGNETIC FIELD NEAR A CURRENT-CARRYING WIRE

In 1819, the Danish physicist H. C. Oersted placed a current-carrying wire near a compass needle during a lecture demonstration before a group of students. Although he predicted that the current would cause a force on the compass needle, the details of the results surprised him. What do you predict will happen? In the activities that follow you will have an opportunity to predict and observe the characteristics of the magnetic field carried by currents. To do the activities in this section you will need:

- 1 lantern battery, 6 V
- 3 alligator clip leads, 20 cm
- 1 microswitch
- 1 rod stand
- 1 aluminum three finger clamp (non-magnetic)
- 2 aluminum support rods (non-magnetic)
- 1 right-angle clamp
- 1 mass, 1 kg (to hold down wire)
- 1 cardboard sheet, 8″ × 8″ (taped to a rod)
- 6 small compasses
- tape

Recommended group size:	4	Interactive demo OK?:	N

27.2.1. Activity: Magnetic Fields from Currents?

a. Do you expect to see a magnetic field in the vicinity of a straight current-carrying wire? Why?

b. If your answer to part a. is yes, do you expect the magnitude of the field to increase, decrease, or stay the same as the distance from the wire increases? Why?

c. Here's a tough one. In which direction do you think the magnetic field will point near the wire? What do you think will happen to the direction of the magnetic field if the direction of the current is reversed?

d. What do you think will happen to the magnitude of the magnetic field if the current is reduced?

Let's repeat some of Oersted's observations and study the pattern of magnetic field lines in a plane perpendicular to a long straight conductor that is carrying current. To do this you will need to set up your equipment as shown in Figure 27.1.

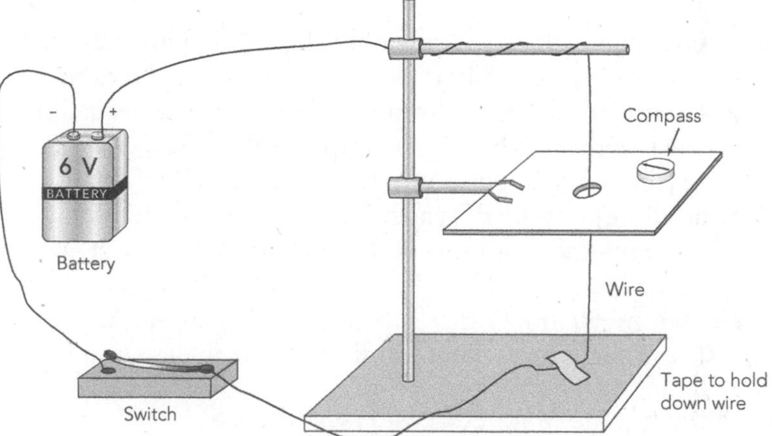

Fig. 27.1. Apparatus for repeating Oersted's observations on the magnetic field produced by a current. The microswitch is wired in a NO (normally open) position.

Wire the battery, switch, and wires in series. The center wire can be poked through a hole in a piece of cardboard with the plane of the cardboard lying perpendicular to the wire. *Turn the current on only when you are making observations;* it saves the batteries.

Check your small compass before starting: the needles sometimes get stuck. Make sure it's pointing north and swinging freely before the current is switched on.

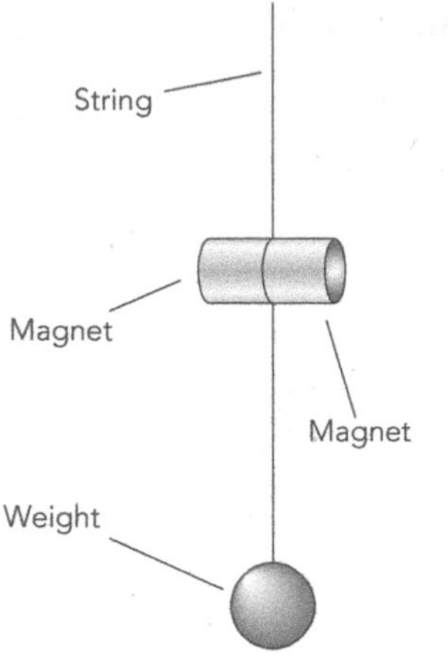

Fig. 27.2. Alternative to compass. A string is clamped between two small, cylindrical neodymium magnets. A small weight is suspended from the string.

Fig. 27.3. A paper arrow is taped to the magnets.

If you prefer, you can use the assembly shown in Figures 27.2 and 27.3 in place of a compass—this assembly is more sensitive than the compass and will show the field at greater distances from the wire. Allow two small neodymium magnets to come together with a string between them. Suspend a small weight from the string (the top of a soda bottle will do nicely). Cut an arrow (approximately 2 inches long) out of stiff paper (such as a manila file folder) and tape it to the magnets, taking care not to get the tape tangled in the string. Hold the assembly by the unweighted end of the string and use it like a compass. (It will help if you attach the arrow so that it points north!) If you use this method, you needn't thread the wire through cardboard resting on a clamp stand.

27.2.2. Activity: The Magnetic Field Near a Wire

a. First, wire the battery so that positive current is passing through the wire from bottom to top. Use the space below to map out the magnetic field directions with arrows. Move the compass slowly in a small imaginary circle centered on the wire and record the direction of the needle. Explain what happens to the direction of the needle as the compass is moved around in the circle.

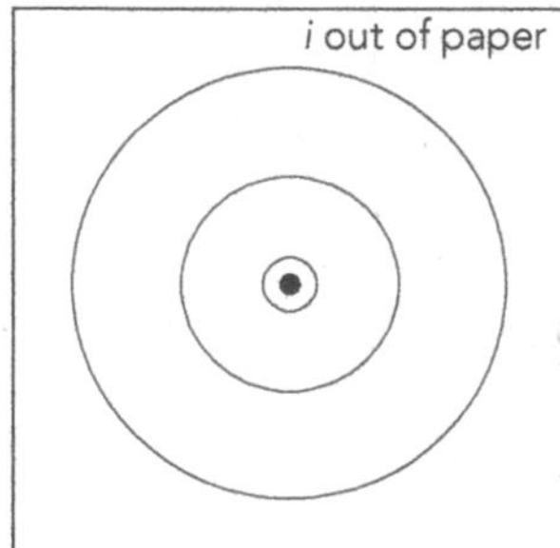

b. Fiddle some more. What happens when the direction of the current is reversed? Wrap either your left or your right hand around the wire with your thumb in the direction of the current, and figure out a rule for predicting the direction of the magnetic field surrounding the wire. **Hint:** (1) Review the definition of magnetic field direction from the last unit and (2) pay attention to the direction of your fingers!

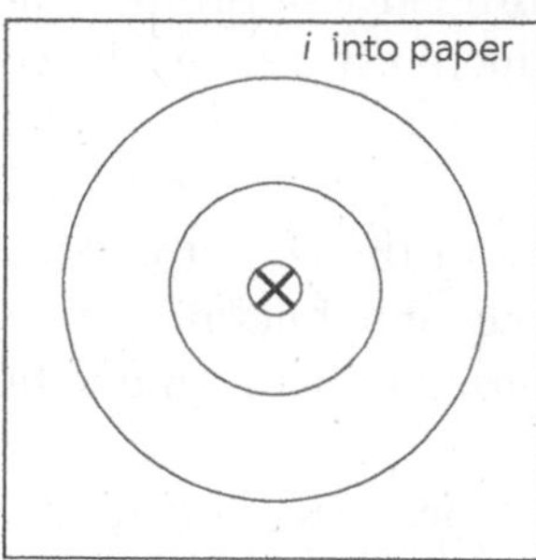

c. What happens as the compass is moved around a circle that is further away from the wire? Does the strength of the magnetic field stay the same? Increase? Decrease?

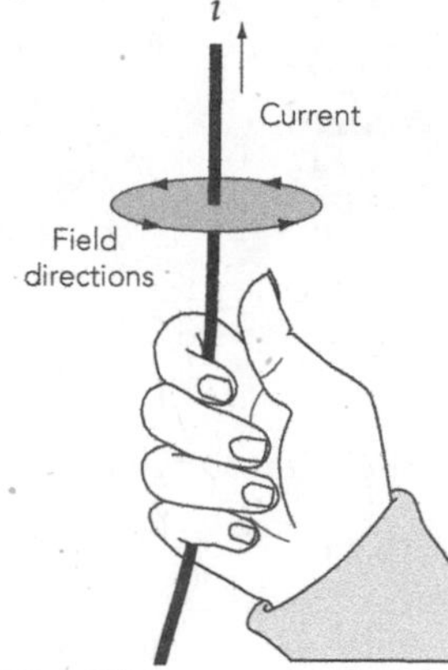

Fig. 27.4.

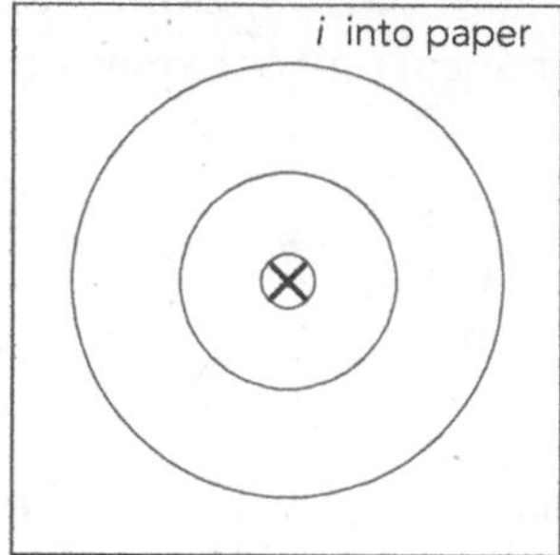

d. Take one of these batteries out of the circuit to reduce the current in the circuit. Hold the switch down for a short amount of time. What happens to the apparent strength of the magnetic field at a given distance when the current in the wire is decreased?

e. How good were the predictions you made in Activity 27.2.1? Did anything surprise you about the actual observation? If so, what?

27.3. DO MAGNETIC FIELD SOURCES SUPERIMPOSE?

You should have established that a current-carrying wire has a magnetic field associated with it. Now, how can we determine the influence of combinations of current-carrying wires? Do the principles of superposition, which work when we combine the electric fields associated with static charges, also work for currents?

Examine the different configurations of the wire in the circuit of Figure 27.5 and predict what the relative strengths of the magnetic field might be at various locations near the circuit. Then you should observe the magnetic field strengths.

To do this observation you will need the same apparatus you used for Activity 27.2.1.

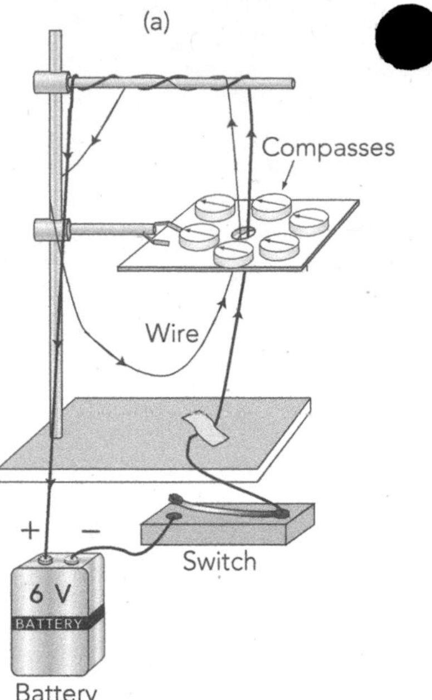

27.3.1. Activity: Magnetic Fields from Different Wiring Arrangements

a. How do you predict the strength of the magnetic field due to two wires carrying current in the same direction (see Fig. 27.5a) will compare to the strength due to one wire carrying the same current? Explain the reasons for your prediction.

b. How do you predict the strength of the magnetic field due to two wires carrying current in opposite directions (shown in Fig. 27.5b) will compare to the strength due to one wire? Explain.

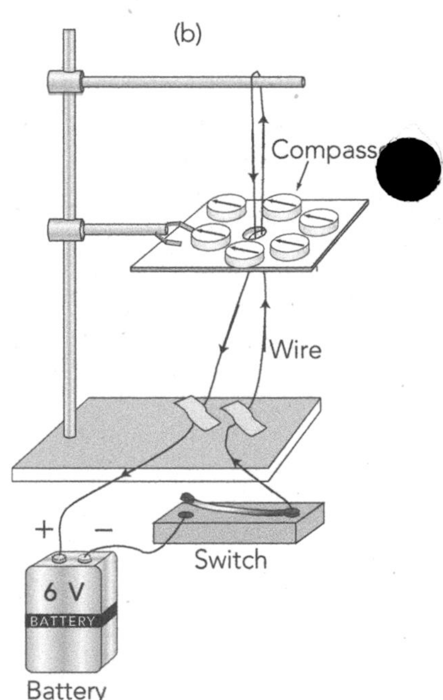

c. Adapt the wire path in Activity 27.2.2 so that two lengths of wire run very close to each other and carry current in the same direction (as shown in Figure 27.5a). Compare the strength of the magnetic field arising from this configuration to that of the field arising from a single wire. Is it weaker, stronger, or the same?

d. Adapt the circuit in Activity 27.2.2 so that two lengths of wire run very close to each other but carry current in the opposite direction (as shown in Figure 27.5b). Compare the strength of the magnetic field arising from this configuration to that of the field arising from a single wire. Is it weaker, stronger, or the same?

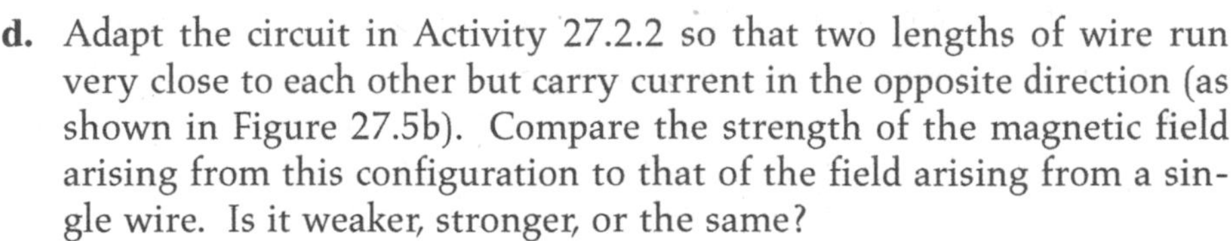

Fig. 27.5. Circuit with different locations of wire sections. (a) Wires paired that carry current in the same direction. (b) Wires paired to carry current in opposite directions.

27.4. AMPERE'S LAW– A MATHEMATICAL EXPRESSION FOR $\vec{B}$

A particularly useful law was proposed by the French physicist André Marie Ampère, who became so excited by Oersted's observations of the magnetic behavior of current-carrying wires that he immediately devoted a great deal of time to making careful observations of electromagnetic phenomena. These observations enabled Ampère to develop his own mathematical equation describing the relationship between current in a wire and the resulting magnetic field produced by the current. Conceptually, Ampère's law is a two-dimensional analog to Gauss' law because it relates the line integral of the magnetic field around a closed loop to the current enclosed by that loop. Ampère's law is given by

$$\oint \vec{B} \cdot d\vec{s} = \mu_0 i^{\text{enc}} \tag{27.1}$$

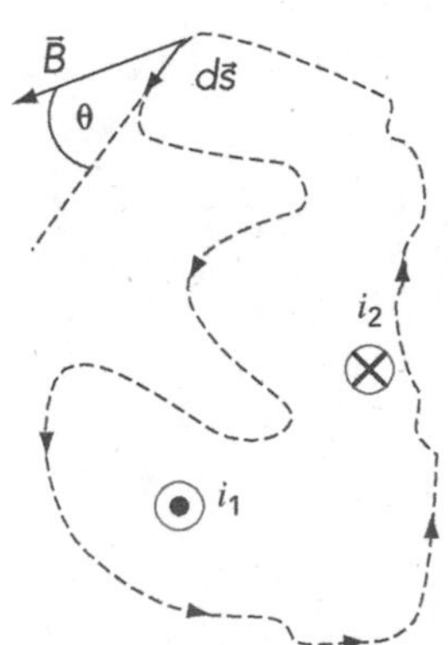

Fig. 27.6. A general Ampèrian loop with net current $i = i_1 - i_2$. The dotted line represents one of an infinite number of possible closed loops that enclose the two wires carrying currents $i = i_1$ and i_2. **Note:** The dotted loop represents an imaginary path, not a wire!

where μ_0 is called the free space magnetic constant (or *permeability*) and i^{enc} is the net electric current passing through the loop. The funny integral sign with the circle in the middle tells the reader that the integral is a line integral around a closed loop. The loop is broken up into an infinite number of little vectors $d\vec{s}$ lying along an arbitrary closed loop that surrounds a current. For each of the $d\vec{s}$ vectors the component of the magnetic field $\vec{B}$ that lies parallel to $d\vec{s}$ is found. Thus, $\vec{B} \cdot d\vec{s} = B \cos\theta \, ds$. Each of these pieces is then added up around a complete loop. This is shown in Figure 27.6.

As is the case with Gauss' law, Ampère's law works best for symmetric geometries. For example, let's use Ampère's law to find the magnetic field caused by a current i flowing through a long cylindrical straight wire. We can draw an imaginary Ampèrian loop around the wire as shown in Figure 27.7.

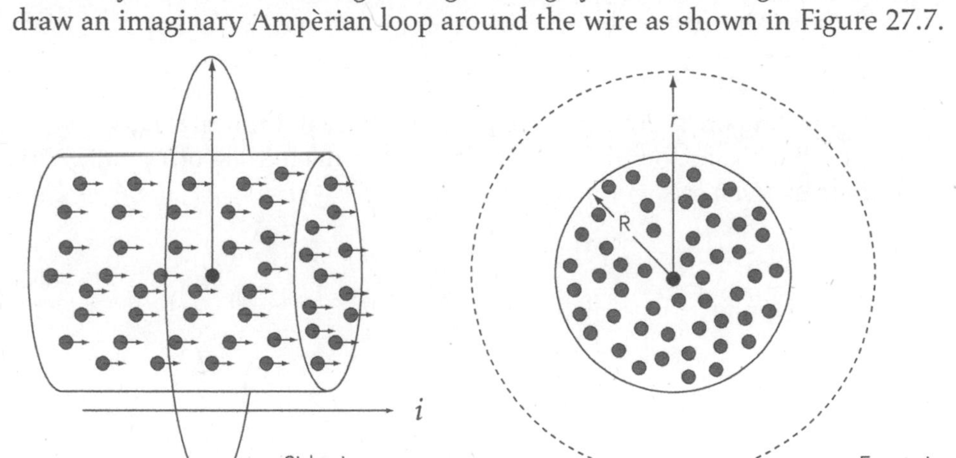

Fig. 27.7. An imaginary circular Ampèrian loop of radius r constructed outside a conductor of radius R which carries a current i. The moving charges are indicated by the small gray dots.

27.4.1. Activity: The Magnetic Field Outside a Wire

a. In Figure 27.7, assume that the current is coming out of the page. Use your previous observations and the right-hand rule to sketch the direction of $\vec{B}$ along the outer circle in the diagram.

b. What is the angle between $\vec{B}$ and $d\vec{s}$ in degrees as you make a complete loop around the circle?

c. Why is the magnitude of $\vec{B}$ the same at all points on the circle?

d. Show mathematically that $\oint \vec{B} \cdot d\vec{s} = 2\pi r B$ for an imaginary circular loop that can be constructed around a straight wire.

e. Using Ampère's law, show mathematically that, for a circular loop outside the conductor (that is, for $r > R$), the magnitude of the magnetic field is given by

$$B = \frac{\mu_0 i^{enc}}{2\pi r}$$

27.5. ARE CURRENT-CARRYING WIRES A HEALTH HAZARD?

There is some indication from epidemiological studies that individuals who live near high-power transmission lines or who make regular use of devices such as electric blankets, heating pads, hair driers, or water beds are at increased risk of developing cancer. It is believed that the biological damage is due to changing magnetic fields associated with the currents carried by wires.

Most of the electrical energy transmitted and used in contemporary homes and industries does not involve steady currents. Instead, the currents in wires typically alternate in direction 60 times each second. Since a current-carrying wire produces a magnetic field, wires carrying 60 hertz alternating currents also have 60 hertz alternating magnetic fields surrounding them. Biologists are conducting laboratory studies that expose single cells, groups of cells, organs, and small animals to low level 60-hertz magnetic fields. There is evidence that weak magnetic fields can interact with receptor molecules on cell surfaces that trigger changes within cells. Such changes include the rates of production of hormones, enzymes, and other proteins.

Although none of the epidemiological studies of cancer rates in human populations or the laboratory studies on animals is conclusive, many scientists are concerned about exposures various people have to magnetic fields.

No study of the health effects of magnetic fields can be conducted without having a way of measuring the magnetic fields associated with various electric devices and of estimating the *doses* of magnetic energy to which people are exposed. Let's take as a case study a hypothetical person who uses an old-style electric blanket during a cold winter. By measuring the maximum magnetic field associated with an old-style electric blanket and making some approximations, we could provide scientists with valuable dose estimates.

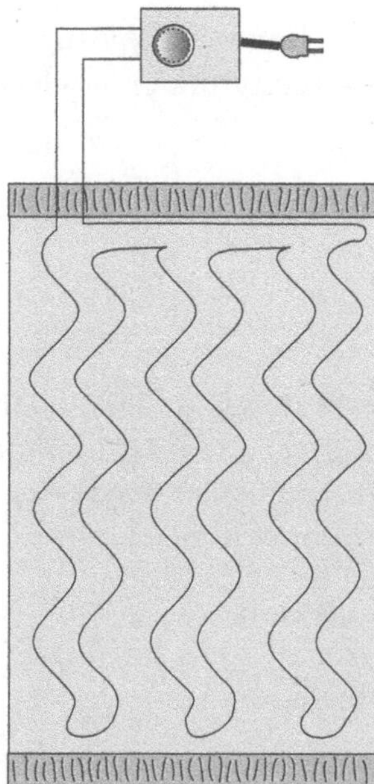

Fig. 27.8. Approximate wiring schematic for some electric blankets.

27.5.1. Activity: Annual B-Field Exposure Estimates for Electric Blanket Users in Pennsylvania

a. If the power rating of a typical electric blanket is 120 watts when it is plugged into a standard household voltage source of 120 volts AC at 60 hertz, what is the average magnitude of the current through the wires in the blanket?

b. What is the average magnetic field in milli-teslas (mT) at a distance of 1 centimeter from a single wire? At 10 centimeters? **Note:** 1000 mT = 1.000 T.

c. *Very approximately,* if a portion of a person's skin is 1.0 cm from a wire, what magnetic field is the skin exposed to?

d. Approximately how many hours a year will a person who uses an electric blanket be sleeping under that blanket? (You can report a range of hours if you like.) Explain your reasoning.

Fig. 27.9.

e. How many milli-tesla-hours of exposure is there to the skin of a person who is sleeping directly under an electric blanket?

You can measure the 60-hertz magnetic field near an electric blanket using a Hall effect sensor. The *Hall effect* and its use in measuring magnetic fields are explained in many introductory physics textbooks. To measure the average B-field near an electric blanket you will need:

- 1 computer data acquisition system
- 1 magnetic field sensor*
- 1 pre–1991 electric blanket

Recommended group size:	2	Interactive demo OK?:	Y

*The magnetic field sensor (MG-BTA) distributed by Vernier Software & Technology and the sensor (CI-6520A) distributed by PASCO Scientific both have sufficient sensitivity to measure fields of < 1 mT.

> **Note:** Since sometime in the early 1990s, a number of electric blanket manufacturers redesigned their products to minimize the magnetic fields surrounding the wires. The following activity will only work properly with older blankets.

Biologists suspect that rapidly changing magnetic fields are potentially harmful, while steady ones such as the Earth's magnetic field are not. Typically, the current passing through a household appliance such as an electric blanket changes its direction 50 or 60 times each second. Thus we are interested in having you measure the difference between the Earth's steady magnetic field and the changing magnetic field of the electric blanket produced by alternating household currents.

If possible, your data-logging software should be set up to display the difference of the magnetic field when it is on, and when it is off. If you use a Vernier magnetic field sensor that consists of an SS94A1 Hall effect sensor attached to an amplifier, you should set the amplifier switch to high amplification (X200). At the X200 amplification the maximum magnetic field in millitesIas is given by

$$\text{magnetic field difference: } B\ [\text{mT}] = \frac{\text{voltage difference}}{.625}$$

where the voltage difference is the difference between the sensor output voltages with the appliance on and off.

27.5.2. Activity: Measuring the B-Field Near an Electric Blanket Wire

a. If a graph of the alternating current through the electric blanket varies as shown in the following diagram, what should the shape of the graph of the magnetic field as a function of time look like? Describe the shape in words and sketch it below.

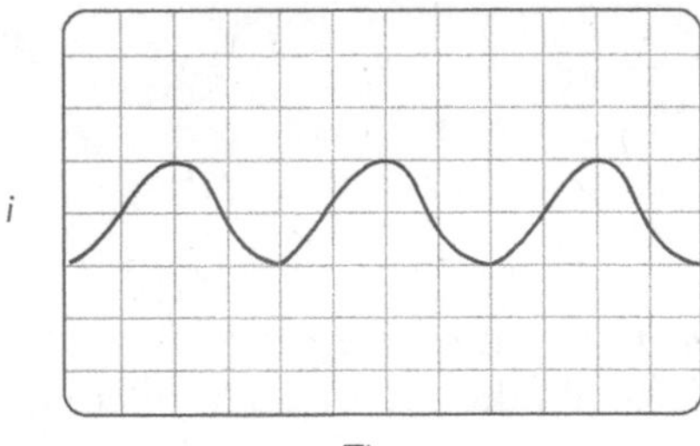

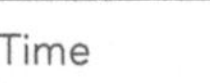

b. Set the voltage logging software to a high data collection rate. We'd suggest taking 4000 data points per second, for one-tenth of a second. Also the magnetic field sensor should be set to maximum sensitivity. The data averaging should be set to "15 points" or more. The magnetic field sensor reacts to a magnetic field that is *perpendicular* to its flat area. Explain how the sensor should be placed relative to the orientation of a wire inside an electric blanket to get the maximum B-field measurement. Draw a sketch, if needed.

c. After some practice with the sensor, set up your data logger software file or use file L270502 to obtain a graph of the magnetic field change as a function of time near a wire in an electric blanket. Start by placing the sensitive area of the Hall effect sensor about 1 cm or less from a stretch of blanket wire. Next zero the sensor with the blanket turned off. *Then, without moving the sensor at all,* turn on the blanket and measure the change in magnetic field for 0.1 seconds.

Fig. 27.10.

d. Use the "analyze" feature of the software to find the period of oscillation of the magnetic field produced by the blanket wire. Then calculate the frequency of the magnetic field change. *Show and explain your calculations.*

e. Use the "analyze" feature to find the maximum change in magnetic field that the sensor detects at a distance of about 1.0 cm from the blanket wire.

f. How does your measured result for the magnetic field at about 1 cm from a blanket wire compare with the value you estimated in Activity 27.5.1b?

g. Suppose you were a blanket manufacturer under pressure to produce a much safer blanket. Use the principle of superposition to design a wiring scheme for your blanket that is safe. Sketch your wiring scheme in the space below. Explain why your design is safer! Write an ad for it if you like.

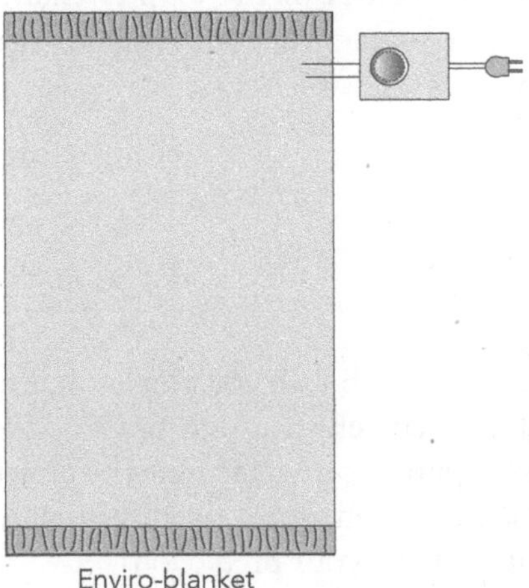
Enviro-blanket

FARADAY'S LAW

27.6. THE MAGNETIC FIELD AT THE CENTER OF A CURRENT LOOP

During the next several sections you will explore some effects of changing magnetic fields. One way to produce a changing magnetic field is by varying the current in a loop of wire. Let's predict and observe the direction and relative magnitude of the magnetic field inside a coil consisting of one or more circular loops of wire as shown in Figures 12.11 and 12.12.

In order to predict the *direction* of the magnetic field at the center of the coil due to the current in one of its loops you can use the rule you devised and explained in Activity 27.2.2b. For the prediction and investigation of the magnetic field in the center of a current loop, you'll need:

- 1 insulated wire, 1 m long (16 AWG w/ thermoplastic insulation)
- 1 large, flat 200-turn coil (known as a "field coil")
- 1 resistor, 2 Ω/2 W
- 1 D-cell battery, 1.5 V, alkaline
- 1 D-cell holder
- 1 microswitch
- 2 alligator clip leads, > 10 cm
- 1 small compass
- 1 computer data acquisition system
- 1 magnetic field sensor*
- 1 ammeter, 1 A
- 1 Lucite holder (or 3″ length of wooden dowel)

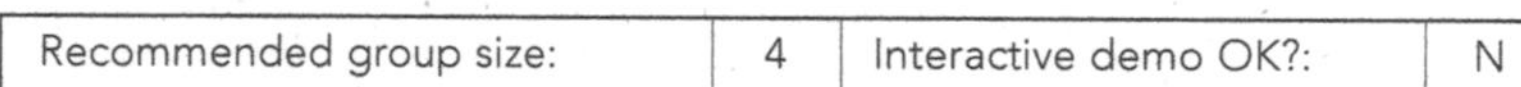

Recommended group size:	4	Interactive demo OK?:	N

Single loop

Fig. 27.11a. A single wire loop carrying a current i.

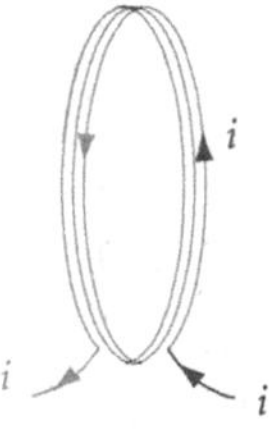

Fig. 27.11b. A multiturn loop carrying current i.

27.6.1. Activity: The Magnetic Field in a Loop

a. On the basis of your observation of the magnetic field surrounding a straight wire, what direction do you think the magnetic field will be in the center of the single loop shown in Figure 27.11 above? How do you expect the magnitude of the field at the center of the loop to change if you make two loops? Three loops? Cite evidence from previous observations to support your prediction.

b. Wrap the wire once around the dowel or Lucite holder, making a single loop, and slip the loop off the dowel. Set up a current through the loop in the direction shown in Figure 27.11. Use a compass to determine the direction of the magnetic field at the center of the loop and sketch the direction in the space on the right. How does it compare with your prediction?

*The field coil should be the one slated for use in Activity 27.11.1.

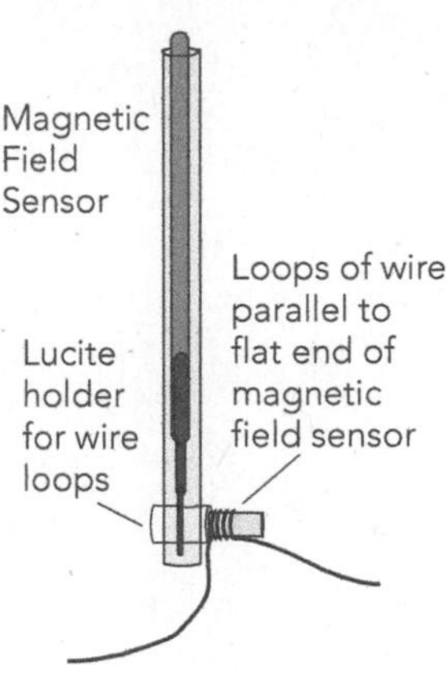

Top view of
Lucite holder

Fig. 27.12. Wire loops and tip of Magnetic Field Sensor. Note that the loop and the flat surface of the Hall probe element of the sensor should be parallel. A small lucite holder can be devised for making this parallel arrangement.

c. To eliminate the effects of the Earth's magnetic field, open the L270601 experiment file and "zero" the magnetic field sensor with the current turned off. Then turn on the current to measure the magnetic field change due to one loop. Next measure the magnitude of the field in milli-teslas [mT] as you coil the insulated wire into more loops. Use the ammeter to measure the current through the wire. Record the measurements in a spreadsheet. Then, graph B vs. N and affix the spreadsheet and graph to the following space. (Your spreadsheet should look roughly like the one shown.) Do your observations agree with your prediction in part a. above? **Note:** (1) Tape the sensor to the table so it does not move as you add loops. (2) DO NOT leave the current in the wire for more than a few seconds at a time or the battery will wear out.

N	i	B_{meas}	B_{loop}
(# of loops)	(A)	(mT)	(mT)
0	—	0	0
1			
2			
3			
4			

Note: For a Vernier magnetic field sensor, $B = \dfrac{V_{sensor}}{.625}\ [mT]$

d. In your circuit, replace the simple wire loop with a 200-turn "field" coil. Measure the current through the coil and the magnetic field at the center of the coil and record these values below. *You will need these values in the next sections.* What is the ratio of B to Ni?

$i_{field}=$ ________ $B_{field}=$ ________ $\dfrac{B}{Ni} =$ ________

____

A very useful result of the formal mathematical calculation for a circular coil of wire is that *the magnetic field at the center of the coil is proportional to the current flowing through its windings and to the number of turns of wire in the coil.* Thus, we will be using the expression

$$B \propto Ni \tag{27.2}$$

in the next sections as we explore Faraday's law.

27.7. MICHAEL FARADAY'S QUEST

In the nineteenth century, the production of a magnetic field by a current-carrying wire was regarded as the creation of magnetism from electricity. This led investigators to a related question. Can magnetism create an electric field capable of causing current to flow in a wire? Michael Faraday, thought by many to be the greatest experimental physicist of the nineteenth century, attempted numerous times to produce electricity from magnetism. He reportedly put a wire that was connected to a galvanometer near a strong magnet, but no current flowed in the wire. Faraday realized that getting current to flow would involve a kind of perpetual motion unless the magnet were to lose some of its magnetism in the process. Although the law of conservation of energy had not yet been formulated, Faraday had an intuitive feeling that the process of placing a wire near a magnet should not lead to the production of electrical current.

Faraday wrestled with this problem off and on for ten years before discovering that he could produce a current in a coil of wire with a *changing* magnetic field. This seemingly small feat has had a profound impact on civilization. Most of the electrical energy that has been produced since the early nineteenth century has been produced by changing magnetic fields. This process has come to be known as *induction.*

To make some qualitative observations on electric field "induction" and associated currents, you'll need:

- 1 galvanometer
- 4 assorted wire coils (with different areas and numbers of turns)
- 2 alligator clip leads
- 1 bar-shaped or rod-shaped magnet
- 1 U-shaped magnet

Recommended group size:	2	Interactive demo OK?:	N

The goal of these observations is two-fold—first, to get a feel for what induction is like, and, second, to discover what factors influence the amount of current induced in the coil. To start your observations you should wire one of the coils in series with a galvanometer and fiddle around with the bar magnet in the vicinity of the coil.

27.7.1. Activity: Current from a Coil and Magnet

a. Play around with the coils and magnet and make a list of as many factors as possible that will determine the maximum current that can be induced in the coil. Each coil has a characteristic resistance and whenever a current is induced in it there is a potential difference created across the coil. This magnetically generated potential difference is called an *electromotive force* or emf—often denoted as $\mathcal{E}$. **Note:** Typically a galvanometer has a "push-to-read" button switch that protects it by shunting current through an equivalent resistor. Be sure to hold this button down while making your observations.

b. Is it possible to have a current or electromotive force in the coil when the magnetic field is not changing in the center of the coil? If necessary, make more observations and explain your results.

This is a good time to make more careful observations on the relationship between various factors that influence the magnitude of the electromotive force induced in a coil. Pick a factor from the list above that can be observed directly and make more detailed observations on its effects. See if you can hypothesize a *simple* mathematical relationship for your factor. For example, you might find that the electromotive force increases with the cross-sectional area of the coil. This could lead you to the intelligent guess (that's what's meant by a hypothesis) that the electromotive force induced in a coil is proportional to its area ($\mathcal{E} \propto A$), and so on.

27.7.2. Activity: Describing an Induced Electromotive Force

a. How do you think the electromotive force induced in a coil depends on the *rate* at which the magnetic field changes in it? **Hint:** Is there any electromotive force induced whenever the magnet and coil are at rest relative to each other?

b. How do you predict the electromotive force induced in a coil depends on the area of the coil?

c. How do you predict the electromotive force induced in a coil depends on the number of turns in the coil?

d. Are there any other factors that you think might influence the electromotive force?

e. Check with some of your classmates and find out what relationships they are hypothesizing for other factors. Write down a trial equation that describes the induction of an electromotive force as a function of the factors you think are important.

27.8. SOME QUALITATIVE OBSERVATIONS OF MAGNETIC INDUCTION

You should be convinced by now that: (1) currents can be induced in a conductor in the presence of a changing magnetic field; and (2) currents cause magnetic fields. Let's observe two phenomena that depend on one or both of these two facts using the following equipment:

- 1 solenoid with an 110 VAC plug
- 1 pickup coil with a small bulb attached (with a larger diameter than the solenoid)
- 1 Lenz's Law Demonstrator

Recommended group size:	All	Interactive demo OK?:	Y

Phenomenon #1: Magnet, Pickup Coil, and Light Bulb

Suppose a 60 hertz alternating current is fed into a solenoid (which consists of a long wire wound into a series of circular wire loops) to create an electromagnet with a changing magnetic field with $dB/dt = A\sin\omega t$. What happens when a coil of wire, with a light bulb attached to it, is placed over the solenoid as shown in Figure 27.13 below?

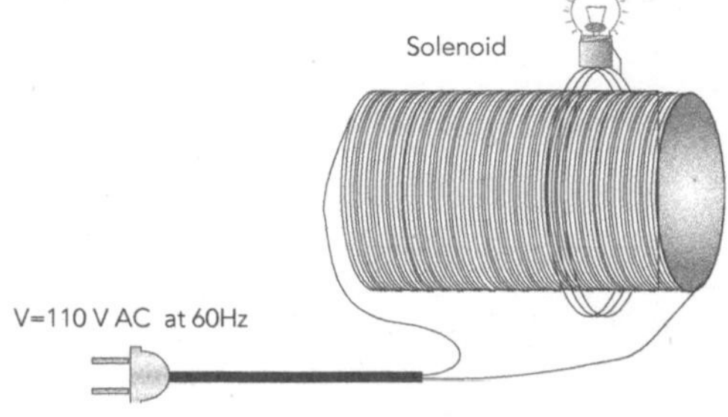

Fig. 27.13. A pickup coil with light bulb attached in series surrounding a solenoid but not touching it.

Phenomenon #2: The Metal Tube and Cylinders

Suppose a non-magnetic cylindrical object is dropped through a metal tube of length L. How fast will it fall? Suppose a cylindrical magnet is dropped through the same metal tube. What might be different? Why?

27.8.1. Activity: Induction Phenomena

a. What did you predict for phenomenon #1, in which a coil with a bulb attached to it surrounds a changing magnetic field?

b. Explain phenomenon #1.

c. What did you predict for phenomenon #2, in which two objects are dropped down a conducting tube? What did you see?

d. Explain phenomenon #2.

27.9. A MATHEMATICAL REPRESENTATION OF FARADAY'S LAW

Let's consider the effect of a magnetic field $\vec{B}$ that is uniform in space at a given moment but changes in time. How much electromotive force, $\mathcal{E}$, will be induced?

By performing a series of quantitative experiments on induction, it can be shown that the electromotive force, $\mathcal{E}$, induced in a coil of wire that is in a changing magnetic field, is given by the equation

$$\mathcal{E} = -N\frac{d\Phi^{m\ ag}}{dt} \quad \text{(Faraday's Law for } N\text{-turn coil)} \qquad (27.3a)$$

where the magnetic flux through a single loop of the coil, Φ^{mag}, is given by the expression

$$\Phi^{mag} = \vec{B}\cdot\vec{A} \qquad (27.3b)$$

where ıis the average magnetic field inside the coil, and $\vec{A}$ is a vector whose magnitude is the cross-sectional area of the coil and whose direction is given by the normal to that cross section. Thus, Faraday's law relating electromotive force to flux can be expressed in terms of magnetic force and area.

$$\mathscr{E} = -N\frac{d\Phi^{mag}}{dt} = -N\frac{d(\vec{B}\cdot\vec{A})}{dt} \qquad (27.3c)$$

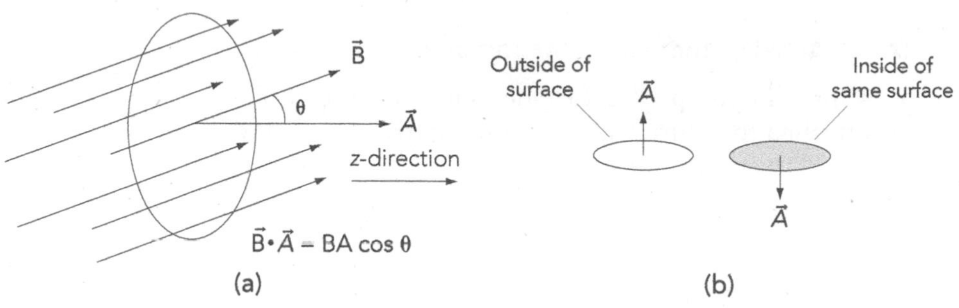

Fig. 27.14. (a) Magnetic flux through an area A is the dot product of the magnetic field vector and the vector normal to the area. (b) The $\vec{A}$ vector points from inside out.

Whenever the plane of a coil of area A is perpendicular to the magnetic field vector, then the dot product can be dropped and Equation 27.3c can be written in terms of vector components along a common axis so that

$$\mathscr{E} = -NA_z\frac{dB_z}{dt} \qquad (27.3d)$$

27.10. COMPUTING THE FLUX IN A COIL AS A FUNCTION OF TIME

Consider a *pickup coil* consisting of a coil of wire with N loops of radius R whose plane is perpendicular to the z-axis as shown in Fig. 27.14. Suppose it is placed perpendicular to a uniform magnetic field 1that varies with time but not location so that

$$B_z = B_z^{max}\sin\omega t$$

where B_z^{max} is a constant representing the maximum magnetic field at the site of the pickup coil. The z-component of the magnetic field varies sinusoidally with time between B_z^{max} and $-B_z^{max}$. Since B_z is a vector component, it can be positive, negative, or zero at different times.

27.10.1. Activity: Applying Faraday's Law to a Wire Loop

a. Use Equation 27.3b to show the equation for the magnetic flux through a single loop of wire is given by $\Phi^{mag} = \pi R^2\, B_z^{max}\sin\omega t$.

b. Use the form of Faraday's law, shown in Equation 27.3d to find an equation for the electromotive force, $\mathscr{E}$, in the pickup coil in terms of N, R, B_z^{max}, ω, and t. **Hint:** The cosine function is involved. Why?

c. The shape of the B_z vs. t graph is shown below for two complete cycles of magnetic field oscillations. Draw a graph showing the shape of the $\mathscr{E}$ vs. t graph for the same time period. Be careful to line the two graphs up properly!

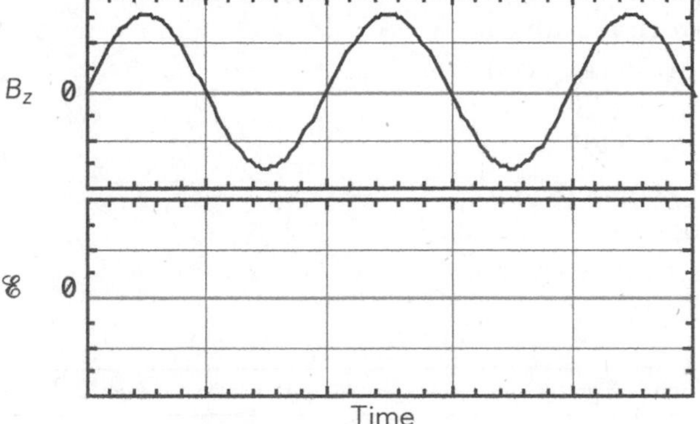

d. Suppose that the magnetic field varies over time in a triangular fashion, as shown in the diagram below. Sketch the shape of the induced electromotive force function in the space below. **Hints:** (1) It is not the same as the electromotive force in part c. (2) Remember that the derivative of a function is its slope at each point in time.

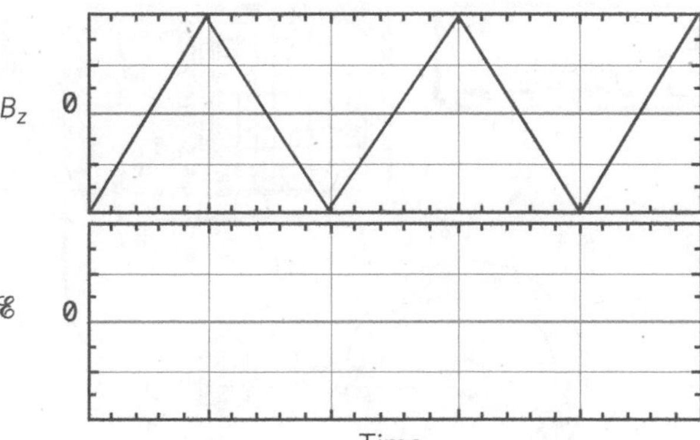

VERIFICATION OF FARADAY'S LAW

27.11. VERIFYING FARADAY'S LAW QUANTITATIVELY

Your mission is to do a quantitative investigation of the electromotive force, $\mathscr{E}$, in an N turn pickup coil as a function of the rate of change of the magnetic flux, Φ^{mag}, through it to see if

$$\mathscr{E} = -N \frac{d\Phi^{mag}}{dt}$$

In this project you can use one current-carrying wire to create a magnetic field that induces an electromotive force in a second coil. The first of these coils, called the *field coil,* can have a changing current from a Wavetek generator pushed through it. The magnetic field that is produced in the center of the field coil also varies with time and is proportional to the current in the field coil. An inner coil, called the *pickup coil,* will have a current induced in it as a result of the time-varying magnetic field. A dual trace oscilloscope can be used to display both the current in the field coil and the electromotive force induced in the pick-up coil. For this activity and the next you will need the following equipment:

- 1 large, flat 200-turn field coil
- 1 2000-turn pickup coil
- 1 400-turn pickup coil
- 1 resistor, 1.2 kΩ
- 1 resistor, 10 kΩ
- 1 signal generator
- 1 oscilloscope
- 1 meter stick or ruler
- 1 protractor

Recommended group size:	3	Interactive demo OK?:	N

The experimental setup pictured in Figure 27.15 can be used to take measurements of the induced electromotive force in the pickup coil as a function either of: (1) the angle between the pickup coil normal vector and the field coil normal vector or (2) the time rate of change of the magnetic flux in the central region of the pickup coil.

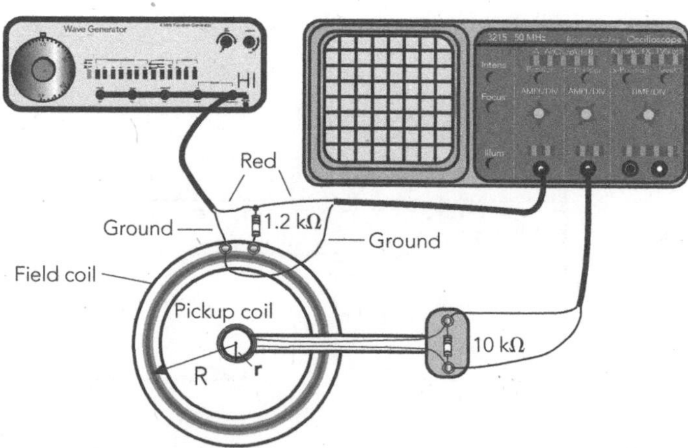

Fig. 27.15. Faraday's law apparatus.

In order to verify Faraday's law you need to use the standard equation for the magnitude of the magnetic field at the center of the field coil as a function of the current in the coil. This equation, which is derived in most introductory physics texts, is given by

$$B_z = \frac{\mu_0 N_f i}{2R} \tag{27.4}$$

where R is the radius of the field coil and N_f is the number of turns in the field coil. If the normal vector of the pickup coil makes an angle θ with the magnetic field vector , then

$$\Phi^{mag} = \vec{B} \cdot \vec{A} = \left[\frac{\mu_0 N_f A \cos\theta}{2R}\right] i \tag{27.5}$$

Thus, for a fixed angle between $\vec{B}$ and $\vec{A}$, the flux, Φ^{mag}, through the pickup coil is directly proportional to the current i. So the change in flux $d\Phi^{mag}/dt$ will also be directly proportional to di/dt. This is important!

Thus, you need to generate a changing magnetic flux in the center of the field coil by changing the current in the field coil. You can then see how the changing flux affects the electromotive force that is induced in the pickup coil.

The first step is to connect the wave generator to the field coil and to the oscilloscope (as shown in Figure 27.15) and put a changing current (in the form of a 1000-hertz triangle wave) into the field coil from the wave generator. **Note:** The voltage drop, ΔV_A, across the 1200 Ω resistor, R_i, can be measured by the oscilloscope from the input A readings for voltage. Ohm's law can then be used to calculate the current, i, in the field coil.

Induced Electromotive Force as a Function of B-field Change

The rate of change of the magnetic flux through the pickup coil is proportional to the rate of change of the current in the field coil. Let's consider a plot of the triangle wave representing the change in current as a function of time as shown in Figure 27.16.

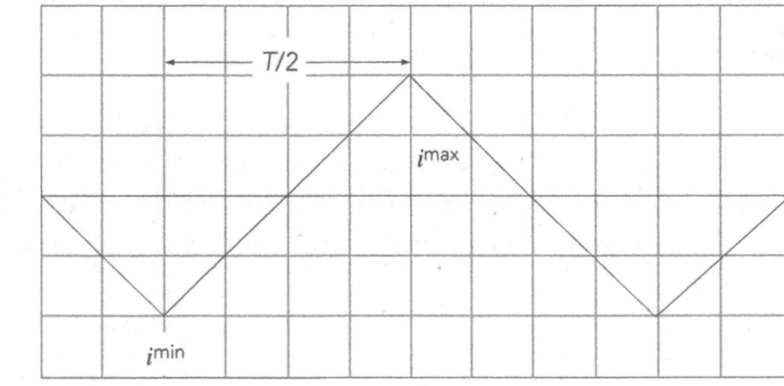

Fig. 27.16. A graph of current vs. time in the field coil when a triangle wave form is fed into a field coil.

It can be seen from the plot in Figure 27.16 that, when the slope of the triangle wave is positive,

$$Slope = \frac{di}{dt} = \frac{i^{max} - i^{min}}{T/2} \qquad (27.6)$$

If the frequency of the wave is set at f on the wave generator, we can use the fact that $T=(1/f)$ to find the slope in terms of f.

$$Slope = \frac{di}{dt} = 2f\,(i^{max} - i^{min})$$

Noting that the negative slope has the same magnitude, in general:

$$Slope = \frac{di}{dt} = \pm 2f\,(i^{max} - i^{min})$$

so that by differentiating Equation 27.5 we can find the rate of change of magnetic flux through each single loop in the pickup coil.

$$\frac{d\Phi^{mag}}{dt} = A_z\frac{dB_z}{dt} = \pm\left[\frac{\mu_0 N_f A}{2R}\right]\cos\theta\ \frac{di}{dt}$$

and finally

$$\frac{d\Phi^{mag}}{dt} = A\frac{dB_z}{dt} = \pm\left[\frac{\mu_0 N_f A}{2R}\right]2f\,(i^{max} - i^{min}) \qquad (27.7)$$

where N_f = the number of turns in the *field* coil

R = the radius of the *field* coil

A = the area of the *pickup* coil

To determine the electromotive force induced in the pickup coil, you should connect the pickup coil in parallel and the 10 kΩ resistor with the oscilloscope in parallel as shown in Figure 27.15.

If Faraday's Law holds, then the magnitude of measured electromotive force induced in the pickup coil should be equal to the calculated value of $N_p d\Phi^{mag}/dt$ where N_p is the number of turns in the pickup coil.

27.11.1. Activity: Results: $\mathscr{E}$ vs. $N_p\, d\,\Phi^{mag}/dt$

a. The measured electromotive force induced in the pickup coil is given by 0.5 ΔV_B where ΔV_B is the difference between the maximum and minimum voltage from the pickup coil as recorded on input B of the oscilloscope. Explain why electromotive force = $0.5\Delta V_B$ rather than ΔV_B.

b. Vary the output frequency of the triangular wave between about 200V and 1000 hertz. Create a data table to record the value of electromotive force as a function of frequency in the space below.

F(Hz)	Measured pickup coil $\mathcal{E} = 0.5\,\Delta V_B$	Calculated $N_p d\Phi^{mag}/dt$
200		
400		
600		
800		
1000		

c. Use the values of I^{max} and I^{min} to calculate $(Nd\Phi^{mag}/dt)$ for each frequency. Then plot the measured values of $\mathcal{E}$ vs. $Nd\Phi^{mag}/dt$. Perform a simple fit to the data. Show a sample calculation in the space below. **Hint:** To find the values of I^{max} and I^{min}, use the data for ΔV_A^{max} and ΔV_A^{min}, the value of the input resistor R_i=1.2kΩ and Ohm's Law. **Note:** *Do not use the measured electromotive force, $\mathcal{E}$, to find $Nd\Phi^{mag}/dt$!*

d. What is the significance of the slope of the graph?

e. Does Faraday's law seem to hold? Explain why or why not.

Flux as a Function of Angle

So far you have concentrated on measuring electromotive force as a function of the wave form that causes a time-varying magnetic field at the site of the pick-up coil. Suppose the normal vector for the pickup coil makes an angle θ with respect to the normal vector of the field coil. If you hold everything else the same, what happens to the maximum electromotive force induced in the pick-up coil?

27.11.2. Activity: Experimental Results: $\mathscr{E}$ vs. θ

a. Draw a graph of the *predicted* maximum electromotive force as a function of the angle between the field coil and the pickup coil in the space below. Do this for angles between 0° and 180° and explain the theory behind your prediction. **Hint:** How does $\Phi^{mag} = \vec{B} \cdot \vec{A}$ depend on the angle? Please label the axes and specify units.

b. Set up an experiment to measure the maximum electromotive force, $\mathscr{E}$, of a changing magnetic field at the site of the pickup coil as a function of angle for at least six angles between 0° and 180°. Explain what you did and create a data table and a computer-generated graph of your results in the space that follows.

c. How did your results compare with your prediction?

27.12. ELECTRICITY AND MAGNETISM – MAXWELL'S EQUATIONS

James Clerk Maxwell, a Scottish physicist, was in his prime when Faraday retired from active teaching and research. He had more of a mathematical bent than Faraday and reformulated many of the basic equations describing electric and magnetic effects into a set of four very famous equations. These equations are shown below in simplified form for situations in which no dielectric or magnetic materials are present.

Gauss' law in electricity: $$\oint \vec{E} \cdot d\vec{A} = \frac{Q}{\varepsilon_0} \quad (27.8)$$

Gauss's law in magnetism: $$\oint \vec{B} \cdot d\vec{A} = 0 \quad (27.9)$$

Faraday's law: $$\oint \vec{E} \cdot d\vec{s} = -\frac{d\Phi^{\text{mag}}}{dt} \quad (27.10)$$

Ampère-Maxwell law: $$\oint \vec{B} \cdot d\vec{s} = \mu_0 i + \mu_0 \varepsilon_0 \frac{d\Phi^{\text{elec}}}{dt} \quad (27.11)$$

If we add the Lorentz force and the Coulomb force on a single charge in an electric and magnetic field

$$\vec{F}^{\text{net}} = \vec{F}^{\text{elec}} + \vec{F}^{\text{mag}} = q\vec{E} + q\vec{v} \times \vec{B} \quad (27.12)$$

to Maxwell's equations, then *we can derive a complete description of all classical electromagnetic interactions* from this set of equations.

Perhaps the most exciting intellectual outcome of Maxwell's equations is their prediction of electromagnetic waves and our eventual understanding of the self-propagating nature of these waves. This picture of electromagnetic wave propagation was not fully appreciated until scientists abandoned the idea that all waves had to propagate through an elastic medium and accepted Einstein's theory of special relativity; these changes occurred in the early part of the twentieth century. To the vast majority of the world's population the practical consequences of Maxwell's formulation assume much more importance than its purely intellectual joys. Richard Feynman, a leading 20th-century physicist, wrote:

> Now we realize that the phenomena of chemical interaction and ultimately of life itself are to be understood in terms of electromagnetism.... The electrical forces, enormous as they are, can also be very tiny, and we can control them and use them in many ways . . . From a long view of the history of mankind—seen from, say, ten thousand years from now—there can be little doubt that the most significant event of the nineteenth century will be judged as Maxwell's discovery of the laws of electrodynamics. The American civil war will pale into provincial insignificance in comparison with this important scientific event of the same decade.

INDEX